AF541597

Seafood Quality Management

NIPA® GENX ELECTRONIC RESOURCES & SOLUTIONS P. LTD.
New Delhi-110 034

About the Authors

Prof. R. Jeya Shakila is a teacher and scientist working at Tamil Nadu D. J. Jayalalithaa Fisheries University, Nagapattinam. She is basically a Fisheries Science Graduate with her Ph.D. in Food Science from Mysore University. Specialized in Fish Post Harvest Technology for over 25 years, she served as FAO International Consultant on Fish Biochemistry and Fish Quality Testing and worked on curriculum development for Marine Science and Technology College in Eritrea and on upgrading the fish quality testing laboratories in Ukraine. She is recipient of NAAS fellowship award from National Agriculture Science Academy and Best Women Scientist award from the Government of Tamil Nadu for her research contributions. She undertook many research projects funded by DST, DBT, CSIR, ICAR, etc. including International collaborate project with Shanghai Ocean University, China. Her major research contribution includes histamine and biogenic amines, waste utilization, marine nutraceuticals and chemical residue monitoring in fish. She possesses more than 200 research articles and published several books. This book chronicles the importance of seafood quality, associated factors, issues relating to different seafood products, its assessment methods, quality management like pre-requisite programs, HACCP, risk assessment, standards, labelling, traceability and waste management.

Dr. G. Jeyasekaran, a Fellow of the National Academy of Agricultural Sciences (FNAAS), and former Director of Research in Tamil Nadu Dr. J. Jayalalithaa Fisheries University has a working experience of more than 35 years. He also served as Dean (Academics) and Dean, College of Fisheries Engineering. He had his Ph.D. research from University of Agricultural Sciences, Mangalore. He visited countries like Australia, Yugoslavia (now Croatia), Morocco, China, USA, Sri Lanka and Cambodia, under different overseas fellowships, travel grants; and as delegation team leader and invited expert. He is a recipient of TNSCST Young Scientist Fellowship (1998), Tamil Nadu Scientist Award (2018), Best Scientist Award (2015), Dr. M.S. Swaminathan Award for Best Indian Fisheries Scientist (2004), Best Fisheries Teacher Award (2003), and Dr. Hiralal Chaudhuri Fisheries Foundation Young Scientist Award (2002). He established a Niche Area of Excellence in Fish Safety and Quality Assurance in the university, besides creating a separate department of Fish Quality Assurance and Management, masters and doctoral academic programs in it. As Director of Research, he signed several MoUs with Overseas Universities from USA, Spain, Australia, Thailand, China, Nepal, and Ethiopia.

His area of interest includes fish safety, quality, authenticity, microbiology, and biotechnology. He first reported the presence of Listeria monocytogenes in Indian seafood. His research group in Australia was the first to report the presence of viable WSSV and YHV in the frozen imported shrimp products. He operated funded research projects receiving grants from ICAR, DBT, SERB-MFPI, DST, and CSIR. He has more than 200 publications (citations 2865, h-index 28), 3 granted and one published patent. He is served as an expert in FSSAI scientific panel, EIA assessment panel, and MPEDA technical monitoring committee. He is instrumental in getting NABL accreditation of the fish quality monitoring and certification laboratory. He also prepared a Policy Paper on Food Safety Strategies for Indian Fisheries Sector as Convener recently for Govt of India. He was also a Senior Consultant to Govt. of India in the National Productivity Council, Ministry of Commerce & Industries, New Delhi for the evaluation of fisheries projects. Presently, he is serving as an Expert in FAO/WHO Joint Working Group on Microbiological Risk Assessment (JEMRA), and Codex India Working Group on Safe Use of Water for Food Processing.

Seafood Quality Management

R. Jeya Shakila
G. Jeyasekaran

NIPA® GENX ELECTRONIC RESOURCES & SOLUTIONS P. LTD.
New Delhi-110 034

NIPA® GENX ELECTRONIC RESOURCES & SOLUTIONS P. LTD.

101,103, Vikas Surya Plaza, CU Block
L.S.C.Market, Pitam Pura, New Delhi-110 034
Ph : +91 11 27341616, 27341717, 27341718
E-mail: newindiapublishingagency@gmail.com
www: www.nipabooks.com

For customer assistance, please contact
Phone: + 91-11-27 34 17 17
Fax: + 91-11-27 34 16 16

Print ISBN: 978-81-19002-61-0

ebook ISBN: 978-81-19002-55-9

Composed and Designed by NIPA®.

Preface

Seafood is an important source of proteins, micronutrients, minerals and polyunsaturated omega-3 fatty acids required by the human population. Globally, the seafood production reached a record of 184.6 million tonnes in 2022, largely due to the growth of aquaculture. The amount of seafood meant for human consumption globally is 20.2 kg per capita. The global trade of seafoods generated around USD 257 billion in 2021. The consumption of seafoods has increased internationally at an average annual rate of 3.0% compared with a population growth rate of 1.6%. Seafood is forecasted to increase by a further 15% by 2030, and it can provide nutritious food requirements for a larger proportion of human population. The value of traded seafoods accounted for 11% of total agricultural trade in 2020. Among the muscle foods, the contribution of seafoods to the export value is 49%. Despite the major progress in processing, refrigeration and transportation facilities, millions of tonnes of seafoods are lost or nutritionally compromised every year. The food loss and waste is a major issue and is the focus of Sustainable Development Goal that aims at halving wastage by 2030. In fisheries and aquaculture sector, it is estimated that up to 35% of the seafood production is either lost or wasted every year due to food safety, quality and authenticity issues.

Safe seafood is a primary requirement of human health. It is a basic human right to have access to safe, nutritious and healthy seafood. To guarantee this right, food regulatory authorities must ensure that available aquatic food meets safety standards. The implementation of Food Safety and Standards Act, and Food Security Act by the Govt. of India bring out some positive impact on the Indian food industries for providing safe food to the consumers. However, the seafoods marketed in India as well as exported to USA, EU, Japan and other countries need several food safety and quality interventions against the background of emerging aquatic foodborne pathogens, and usage of chemicals like formalin, ammonia, antibiotics, pesticides, heavy metal contaminants, and food additives in seafoods. The Joint FAO/WHO body, Codex Alimentarius Commission, has been at the forefront of bringing out global changes in seafood safety regulations. Similarly, the WHO, FAO, UNEP and WOAH have jointly considered food safety as part of their One Health Plan.

The food regulatory authorities need specialized technical manpower to oversee the implementation of food safety management systems in seafood processing establishments, and technically qualified people to ensure the supply of safe and quality seafood to consumers by the producers and retailers. With the in-depth experience of more than 30 years in teaching and research in the field of seafood safety and quality by means of offering an unique and innovative PG and Ph.D. degree programmes in Fish Quality Assurance and Management at the national level, and operating several research programmes to the worth of crores in the same field, we, the authors, have already published two books for the fisheries profession on seafood safety and quality and this is the third book in the series. The reader of this Textbook not only gain in-depth knowledge on the basics of seafood quality and safety management and but also serves as a handbook for reference. We are glad that the publishers had identified this area of advanced and emerging science, and brought it for the scientific fraternity.

Authors

Contents

1

Quality Dimensions of Seafood

About the Chapter

In this chapter, the importance of seafood in terms of their nutrition, and different methods of processing are dealt. Seafood quality and safety terminologies are explained briefly. The quality parameters affected by rigor mortis in raw, iced and cooked fish are given. Phenomenon of gaping described with reference to rigor mortis. Quality dimension of seafood is explained with intrinsic and extrinsic parameters. Various intrinsic factors affecting seafood quality such as fish species, seasonal changes, parasite infestation and capture conditions are discussed in detail. The next part elucidates the sensory parameters affecting the seafood quality describing the objective and subjective method of quality assessments. Objective method covers both descriptive and discriminative tests. The final part mentions the Codex guidelines available for the conduction of seafood sensory assessment. At the end, the reader gains the knowledge on the basic information about seafood, role of rigor mortis in affecting seafood quality and the sensory quality assessment methods.

Objective

To learn the nutritional importance of seafood and their processing; rigor mortis and their role in seafood quality, intrinsic and extrinsic factors affecting seafood quality and sensory quality assessment methods.

1.1. Importance of Seafood

Seafood has been traditionally a part of diet in many parts of the world. In some countries like India, it constitutes the main supply of animal protein among coastal population. Seafood is a low-calorie food and easily digestible compared to other protein-rich meat and poultry products and hence, it is a good choice for elderly persons to obtain their daily protein requirements. Seafood has high-quality quality enriched with all the essential amino acids required for human health. It provides about 1/3rd of the average recommended daily allowance (RDA) of protein.

Seafood is a low-fat diet with less total fat and saturated fat. Lean fish contain less than 5% total fat, and high fat fish such as mackerel, sardines, tuna, herring, and king salmon, have about 15% total fat. One serving size (i.e. 3 ounce or 85g) of lean fish provides approximately 100 calories, and that of fatty fish gives 200 calories. A large proportion of fat in seafood is unsaturated, mainly polyunsaturated omega-3 fatty acids (PUFA). Omega-3 fatty acids provide various health benefits, mainly they reduce the risk of heart disease, decrease blood fats and cholesterol, and have effects in controlling cancer, arthritis, and asthma. Fatty fish contains high omega-3 fatty acids than lean fish.

Seafood is relatively rich in vitamins, particularly B vitamins. Fatty fish are good sources of Vitamin A and D. Most seafoods have reasonable quantities of phosphorus, potassium and selenium. Shellfish such as clams and oysters are a good source of iron, zinc, magnesium, copper, iodine and other trace minerals. Considering the rich nutritional aspects of seafood, people are Now-a-days turning into fish eaters, as fish meat forms a healthy alternative to red meat.

1.1.1. Processing of Seafood

Seafood are produced either through capture fisheries or though aquaculture. They are processed in many ways, as fresh or frozen or processed forms. Seafood are served at home and in restaurants after culinary preparation. Raw seafood dishes include sashimi, fried, grilled, broiled, baked, barbecued, marinated, or boiled in seasoned liquids. Industrially processed seafoods are salted, dried, smoked, marinated, or canned products. Despite the consumption of seafood in various forms, quality becomes an important criterion for acceptance.

1.2. Seafood Quality

The term "**quality**" refers to the aesthetic appearance and freshness of fish or sometimes refers to the degree of spoilage in fish. The quality also involves "**safety**" aspects such as the presence of harmful bacteria, parasites or chemicals in fish. The term "**quality**" implies different meaning to different people. i.e. for connoisseurs, who select rare expensive fish, crustaceans, and mollusks for their cuisine, the quality is different; for needy persons, who eat cooked, dried, salted, or fermented fish as a supplement to the staple cereals in their daily diet, it is different. The best quality fish has to be consumed within the first few hours of post-mortem. The processors prefer slightly older fish i.e. post-rigor fish, as it is difficult to make fillet to fillet or skin the fish, which is in rigor mortis.

1.2.1. Rigor mortis

The most dramatic change after harvest or capture of fish is onset of rigor mortis. The muscle totally relaxes immediately after death of fish giving the limp elastic texture that persists for some hours. This is **pre-rigor stage** (Fig 1.1). The muscle thereafter contracts, becomes hard and stiff turning the body inflexible to a state of rigor mortis, which lasts for a day or more. This is **in-rigor** stage (Fig 1.1). The resolution of rigor mortis relaxes the muscle again to become limp, but it is no longer elastic as it is before rigor. This is **post-rigor** stage (Fig 1.1). The rate at which onset and resolution of rigor occur varies from species to species. It is influenced by temperature, handling, size and physical condition of the fish.

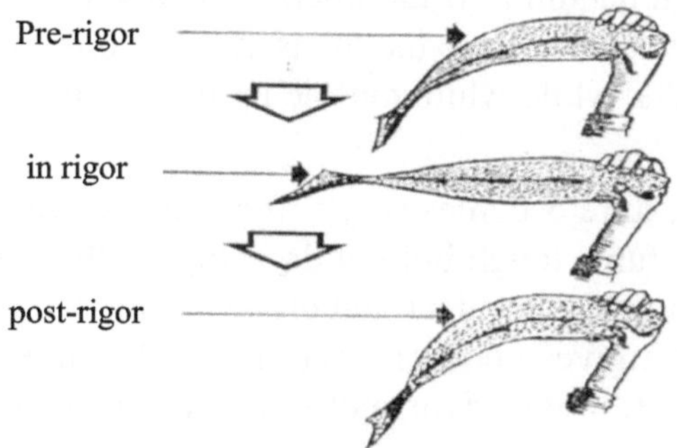

Fig. 1.1 : Stages of Rigor mortis

(Source: G.D. Stroud, Torry Research Station, Aberdeen, UK)

In a temperate fish (e.g. cod), high temperatures give a fast onset and a very strong rigor mortis leading to a condition known as "**gaping**", i.e. weakening of the connective tissue causing a rupture of the fillet. Fig 1.2 shows the gaping in fillet of cod, saithe and witch flounder. On the contrary, in tropical fish, the onset of rigor is accelerated at low temperature i.e. 0°C as compared to 10°C. The onset of rigor mortis in fish thus depends on the difference in sea temperature and storage temperature. When the difference in temperature is large, the time from death to onset of rigor is short and *vice versa.*

If the fish is starved due to depletion of glycogen reserves, or if the fish is stressed, rigor mortis starts immediately after death. The method of stunning and killing the fish also influences the onset of rigor. Stunning and killing by hypothermia accelerates the onset of rigor, while a blow on the head gives a delay of up to 18 hours.

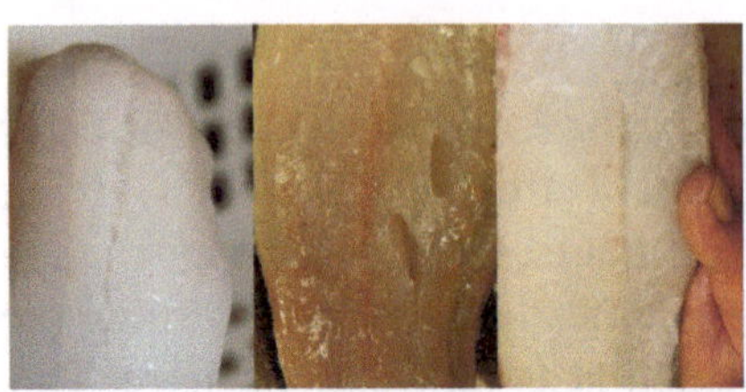

Fig. 1.2 : Gaping in cod (left), saithe (middle), and witch flounder (right) fillets. (Source: Karlsen et al., 2015)

The technological significance of rigor mortis is of much importance when the fish is filleted before or in rigor. If the fish is in rigor, the filleting yield is very poor, and gaping occurs due to rough handling. If the filleting is done in pre-rigor fish, the muscle contracts freely and shortens the fillets after the onset of rigor. Dark muscle shrinks up to 52%, while white muscle up to 15% of the original length.

In the fish cooked in pre-rigor, the texture becomes very soft and pasty, while in those cooked in rigor, the texture turns tough but not dry. Only in the fish cooked post-rigor, the flesh remains firm, succulent and elastic. If the whole fish and fillets are frozen pre-rigor, it gives good product only if thawing is done carefully at a low temperature to allow rigor mortis to pass in frozen condition itself.

1.3. Quality Dimensions of Seafood

Quality dimensions of seafood are measurable attributes inherent to the species. The quality attributes comprise of :

1. Sensory expectations of consumers
2. Health and safety aspects
3. Nutritional value
4. Functional properties
5. Conformation to product standards (i.e. with regard to composition, proportions of components, appearance, ingredients, packaging, labeling, and shelf-life).

These attributes can be quantitatively determined and the measured values are used for assessing the quality grade of seafood. Many factors affect the sensory quality of seafood. Some are related to unprocessed or live fish, while others pertain to processed fish that have passed through seafood production chain.

The factors influencing the sensory quality of unprocessed or live fish such as species, age, sexual maturation, genetics, season, sex, farmed or wild and other growing condition such as water quality, feed and presence of disease or parasites are collectively known as **"Intrinsic parameters"** that affect the seafood quality.

Seafood production chain involves various steps from fish catch to consumption such as catching, slaughtering, cutting, processing, transport and storage. Different steps take place in different locations such as fishing vessels, aquaculture ponds, slaughterhouses, different means of transport, processing industry, fish shops, supermarkets, catering businesses and consumers' homes. Time and temperature, and packaging atmospheres influence the sensory quality in all these steps. Microbial contamination or physical damage further reduces the sensory quality. Quality of the gutting, filleting or other types of pre-processing also affects the sensory quality of the end product. All these factors are collectively termed as "**extrinsic parameters**" that affect the seafood quality.

1.3.1. Intrinsic Parameters

The initial quality of freshly harvested unprocessed or live fish depends on the various attributes such as:

i. Species characteristics

ii. Seasonal biological changes in fish

iii. Parasite infestation

iv. Conditions applied in aquaculture and in fishing

i. Species characteristics

The species-related characteristics are size, appearance and shape, yield of edible parts, content and distribution of dark muscles, flavor and texture of the meat, biochemical composition (predominantly proteins, lipids, and vitamins), the activity of enzymes, and the presence of natural toxins. The size of the fish affects the vulnerability of the catch to damage during handling and processing. The appearance and shape influence the appeal of the fish to the buyer, while the yield of edible parts has an impact on the economy of processing. The color, flavor, and texture of the meat are important factors affecting the suitability of the fish for different methods of culinary preparation and for use in the food industry.

The protein, lipid, vitamin, and mineral components affect the sensory properties of the product and its nutritional value. The enzyme activity influences the susceptibility of the catch to storage changes, such as **rapid post-mortem softening** of fish or squid possessing active proteolytic enzymes, or **freeze**

denaturation in proteins of frozen stored fish having more trimethylamine oxide (TMAO) demethylase. The enzyme activity also has an impact on the suitability of raw material for producing different kinds of products, such as mild salted maatjes herrings or marinades. Non-protein nitrogenous (NPN) compounds affect the sensory properties, safety, and suitability of the fish for different end uses. For instance, TMAO is involved in causing undesirable greening of canned albacore, and biogenic amines, which can cause health hazards. Endogenous toxins make the fish meat of several species unfit for human consumption.

ii. Seasonal biological changes in fish

The seasonal biological changes are responsible for fluctuations in protein and lipid contents in fish muscles, and for the activity of endogenous proteases. The variation in biochemical composition is related to feeding condition, migratory behavior and sexual changes during spawning. Migratory fish has less protein and lipid reserves. Heavily fed fish has high protein content in the muscle tissue. The biological condition of the fish also affects the gel-forming ability of the muscle proteins. In migratory salmonid fishes during spawning, the flesh has a paste-like appearance because of enzymatic degradation of the muscle fiber integrity.

iii. Parasite infestation

Infestation with parasites decrease the technological value of the catch, or constitute severe health hazards for the consumers of raw fish and aquatic invertebrates of various species. Fish get infested with parasites when they feed on intermediate hosts. There are three kinds of fish parasites of public health significance, namely:

a. Round worms (nematodes)

b. Flukes or flatworms (trematodes)

c. Tapeworms (cestodes).

Round worms

The most common round worms that infect fish and invertebrates are *Anisakis* spp., *Pseudoterranova* spp., *Phocascaris* spp. and *Contracaecum* spp, all belonging to the family, Anisakidae. The round worms are associated with salt water fish. The definite host of round worms are marine mammals such as dolphin, seals and seabirds. Eggs shed in their faeces hatch out into larvae and taken up by fish. The larvae bore through intestinal wall and enter the muscle tissue. Consumption of fish with round worms causes Anisakiasis and symptoms are abdominal pain, nausea, vomiting, diarrhea, blood and mucus in stool, mild fever. Allergic reactions with rash and itching and anaphylaxis also occur. *Anisakis simplex* is the only roundworm that causes clinical allergic response.

Flatworms or Flukes

The most common flukes are liver flukes belonging to the family, Opisthorchiidae and some are intestinal flukes of family, Heterophyidae and Echinostomatidae. Fish flukes are found in fresh water fish of both temperate and warm waters. The definite host of fluke are land mammals like dogs and cats. Eggs shed in their faeces are ingested by snails and develop into cercariae larva before released into the water. Larvae penetrate beneath the scales of fish and turn into cysts in the fish tissue. Consumption of fish with fluke causes 'trematodosis', which has asymptomatic early and light infection. If the load is high, malaise is common and severe pain occurs in abdominal region. Chronic infection causes severe morbidity. Clonorchiasis and Opisothorchiasis cause inflammation and fibrosis of adjacent tissues, if adult worms enter small bile ducts of liver.

Tapeworms

The most common tapeworm that infect fish is *Diphyllobothrium*. They are mostly associated with fresh water fish in cold waters. The definite host of tapeworm are fish eating mammals and bird. Eggs shed into water, hatch out into motile embryos, which are taken by copepods to develop into first stage larvae. Fish consume copepods and larvae enter the tissue through walls of digestive tract and develop into second stage larvae. Consumption of fish with tapeworm cause 'diphyllobothriasis', which is long-lasting asymptomatic infection causing abdominal discomfort, diarrhea, vomiting and weight loss. Massive infection causes intestinal obstruction. Vitamin B12 deficiency with pernicious anemia occurs.

The parasites, in general survive in fish tissue for long periods and sometimes, remain inactive for several years. However, fish parasites can be killed by freezing and heating treatments. Regulation (EC) No. 853/2004 places a requirement that fishery products shall undergo a treatment sufficient to kill viable parasites that cause a health hazard to the consumer. Commission Regulation (EU) No. 1276/2011 details that food business operators (FBOs) placing fishery products intended to be consumed raw or marinated, salted or other treated ones must ensure that the product undergo a freezing treatment to kill viable parasites. To kill parasites other than trematodes, the freezing must be carried out at -20°C for not less than 24 h or at -35°C for not less than 15 h. Heating at >60ºC for at least 1 min also destroys parasites. More specifically, fluke larvae of *Opisthorchis* spp. and *Clonorchis* spp. are killed by freezing at -10ºC for 5 days. Heating at temperature of 70ºC for 30 min also kills *Opisthorchis* spp. and *Clonorchis* spp.

iv. Conditions applied in aquaculture and in fishing

The conditions applied in aquaculture and fishing operations affect the technological suitability of the catch. Pre-harvest conditions or the amount of stress placed on live fish during handling also affect quality. Natural and man-made compounds cause undesirable flavour attributes in freshwater fish. Geosmin and 2-methyl-iso-borneal (MIB) are two naturally compounds that cause muddy/earthy flavor in freshwater fish. Geosmin is an organic compound produced by cyanobacteria and actinomyces present in freshwater aquaculture system. When the microbes die and decay, it is released into water. It is a potent environmental contaminant in catfish and carp. It is fat soluble and hence concentrates in the skin and dark muscle tissue of fish. The use of lemon juice helps to reduce the muddy odour, as it gets broken down under acidic condition. The human threshold for geosmin is 15 ppb. MIB is an organic compound responsible for musty flavor in fish. It is produced by cyanobacteria in summer months. The human threshold limit is 35 ppb. Geosmin and MIB is taken from water by fish primarily through the gills.

The quality attributes of fish and fish products change on the way to the consumer site because of time and the impact of post-harvest treatment, sanitary conditions during handling, storage, and processing, and the procedures and materials used for preservation.

1.4. Sensory Parameters

Sensory attributes affect the quality of seafood. Sensory changes in seafood are perceived by using human senses, i.e. appearance, odour, texture and taste. The first sensory changes that occur in fish during storage are with appearance and texture. The characteristic taste of the species develops in the first couple of days during storage in ice.

The sensory characteristic of fish changes depending on the storage method. The appearance of fish stored under chilled condition without ice does not change, but spoils more rapidly than iced fish and hence, an evaluation of cooked flavor is necessary. A characteristic pattern of the deterioration of fish stored in ice is divided into the following four phases:

Phase 1: Fish is very fresh. It has a sweet, seaweedy and delicate taste. The taste can be very slightly metallic. In lean fish, the sweet taste is maximum for 2-3 days after catching.

Phase 2: There is a loss of the characteristic odour and taste. The flesh becomes neutral but has no off-flavors. The texture is still pleasant.

Phase 3: There is sign of spoilage and a range of volatile, unpleasant-smelling substances is produced depending on the fish species and type of spoilage (aerobic, anaerobic). Trimethylamine (TMA) derived from the bacterial reduction of trimethyl-aminoxide (TMAO) has a very characteristic "fishy" smell. In the early stages, off-flavor is slightly sour, fruity and slightly bitter, especially in fatty fish, and in the later stages, it is sickly sweet, cabbage-like, ammoniacal, sulfurous and rancid. The texture becomes either soft and watery or tough and dry.

Phase 4: The fish is characterized as spoiled and putrid.

1.4.1. Sensory assessment

Sensory assessment of seafood is done by using one or more of the five senses to judge the quality of fish. Sight, smell, taste, touch and hearing are the five senses. The term "**organoleptic**" refers to the characteristics of a food as perceived by the senses. Sensory assessment is defined as the scientific discipline used to evoke, measure, analyze and interpret the reaction of different food characteristics as perceived through the senses.

The sensory assessment of raw fish in markets and landing sites is done mainly by assessing the appearance, texture and odor by using human senses. The evaluation process involves three steps:

1. Detection of stimulus by the human sense organs
2. Evaluation and interpretation by a mental process
3. Response of the assessor to the stimuli

While assessing the 'quality', eating quality must be taken as the most important component.

Sometimes an assessment is based on the use of only one sense. For instance, the sensory assessment of a whole whiting or haddock is done by **sight** alone. In some occasions, two or more senses are used, mainly **smell or taste**, in addition to sight. Variations in the sensory response occur in several circumstances to differentiate fish based on **color** due to color blindness, sensitivity to chemical stimuli, response to rancid and cold-storage flavor, pre-or post-rigor fish, etc.

Sensory assessment is the most preferred method over chemical or physical or microbiological instrumental methods, if done under careful controlled conditions by reducing the effects of test environment and personal bias. Sensory assessment is classified into two types, namely **"objective"** and **"subjective"** assessment.

1.4.1.1. Objective assessment

The objective assessment is more "**quantitative**" in nature. The effects of personal influences like bias and feelings are avoided by providing an accurate description of the particular aspect of quality for evaluation by the assessors. e.g. seaweed taste, size grade 3, very soft product, straw-color product, etc. The objective assessment is used for two main purposes.

1. To describe particular aspects of a product quality after production to meet market requirements.
2. To distinguish between two or more products during sorting by eye size or visually inspecting for spoilage or defects.

The objective assessment is preferred to establish, control or assure standards of quality in fish businesses. The objective methods are classified into two groups: **discriminative tests** and **descriptive tests.**

I. Discriminative tests

In discriminative test, the difference existing between two samples is determined. e.g. triangle test and ranking test.

a. Triangle test

This is the most commonly used ISO 4120:1980 method to find out the detectable difference between two samples. The assessors are presented with three coded samples, preferably in duplicates and are asked to identify the two identical and one different sample. The samples are marked A and B and presented in six different ways: ABB, BBA, AAB, BAB, ABA, BAA. The number of panel members should not be less than 12. The results are analyzed by comparing the number of correct identifications using a statistical chart. If the number of responses is 12, there must be 9 correct responses to achieve a significant result at 1% level. Triangle test is used to determine the ingredient substitution in a product, and to select assessors to a taste panel.

b. Ranking test

In ranking test, the sensory panel is asked to arrange the number of samples based on the degree of some specified characteristics e.g. saltiness. Ranking is done quickly than other evaluation methods and hence, used for preliminary screening. It is not suitable for find out individual differences among samples.

II. Descriptive tests

In descriptive test, the difference in the nature and intensity of the samples is determined. e.g. profiling, structured and quality index tests.

a. Profiling

It is a simple method used to assess as single attribute of texture, flavor or appearance It generates a complete description of a sample e.g. flavor profiling. Quantitative Descriptive Analysis (QDA) provides a detailed description of all flavor characteristics e.g. Lyngby developed a QDA method for fish oil. The descriptive terms used are painty, nutty, grassy, metallic to describe the oil. Multivariate analysis correlates single attribute to oxidative deterioration, which is reported in "spiders web" (Fig. 1.3). Profiling can be used for all kinds of fishery products, including fresh fish. A similar method is also used for texture.

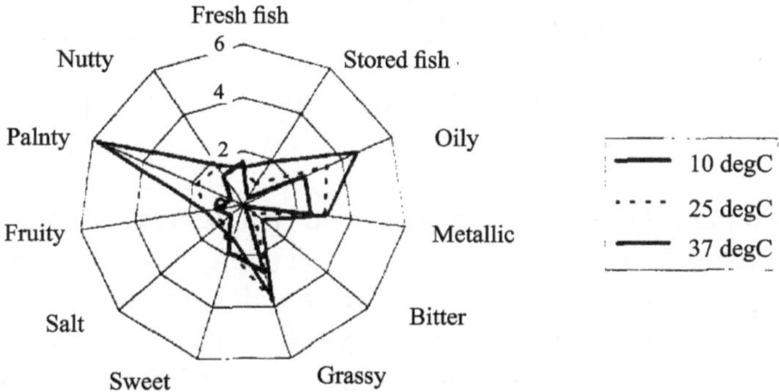

Fig. 1.3 : Spiders web indicating flavor profile of a fish oil on storage at three temperatures (Source : FAO, 1995)

b. Structured scaling method

This method is used to determine the quality and shelf life of fish. The panelists are given with an actual scale showing several degrees of intensity, based on detailed attributes that are given in descriptive words. Objective terms preferred over subjective terms. Standards are included at various points of the scale.

A 10-point scale that describes overall impression of odor, flavor, and texture of cooked fish was developed by Torry Research Station, UK (Table 1.1). The scale is numbered from 0 to 10, 10 indicates absolute freshness, 8 for good quality and 6 for neutral tasteless fish. The limit of acceptability is 4. Using the scale, the graph becomes S-shaped indicating a fast degradation of the fish during phase 1, a slower rate in phase 2 and 3 and finally a high rate when the fish is spoiled.

Table 1.1 : Evaluation of cooked fish

		Grade		Score
Acceptable	No off-odour/ flavour	I	Odour/flavour characteristic of species, very fresh, seaweedy	10 9 8
			Loss of odour/flavour	7
			Neutral	6
	Slight off-odour/flavour	II	Slight off-odours/flavours such as mousy, garlic, bready, sour, fruity, rancid	5 4
Limit of acceptability				
Reject	Severe off-odour/flavour	III	Strong off-odours/flavours such as stale cabbage, NH_3, H_2S or sulphides	3 2 1

The Quality Index Method (QIM) is also a structured scaling method, which assess the quality based on certain attributes of raw fish (skin, eyes, gills, etc.) using a **point scoring system** from 0 to 3. The scores for all attributes are then added to get an overall sensory score, known as "**quality index**". The quality index increases linearly with the keeping time in ice. The total demerit score is used to predict the remaining shelf life. The description of the evaluation of each parameter is written in a guideline. Minor differences in results for any one criterion do not unduly influence the total QIM score. The lower the score the fresher the fish.

c. EU grading scheme

The EU grading scheme is developed for sensory quality assessment of fish in the inspection service and in the fishing industry in Europe. EU regulation No. 2406/96 provides only one type of sensory assessment for seafood, i.e. the EU-quality grading scheme for fresh fish. This method is recommended at the first point of sale to check freshness based on quality attributes. Council decision No. 103/76 introduced this system, which has three quality levels, E (extra), A, B, where E is the highest quality and below B is the level for discarding. There are different schemes applicable for whitefish, bluefish, elasmobranchs, cephalopods and crustaceans. This method is used by marketing personnel or by the Competent Authority.

1.4.1.2. Subjective assessment

The subjective assessment is also termed as "**affective**" test, as it is based on a measure of preference or acceptance. Natural feelings of liking, pleasure, acceptance etc. are freely expressed by assessors e.g. I don't like this fish taste, I prefer product A than product B, this fish is delicious, I prefer this product, etc. It involves expressions of pleasure or degrees of it, and hence termed as "**hedonic**" assessment. It is used in product development or market research. Hedonic responses of consumers to fish products in shops and eating places are the main factors in marketing of products by the industry e.g. consumer surveys.

1.4.1.3. Guidelines for conducting sensory assessment

Codex Alimentarius Commission have given guidelines for the conduction of sensory assessment of fish and shell fish in laboratories (CAG/GL 31-1999). It describes the facilities required and the procedures required for conducting sensory evaluation, and the training of assessors. The sensory characteristics of raw whole gutted or ungutted fish; raw fillets; cooked fillets; frozen vertebrate fish; raw, cooked and frozen crustacean shellfish; and fresh or refrigerated cephalopods are given for ready references.

References

Connell, J.J. 1989. Sensory assessment of fish quality. Torry Research Station Advisory Note No. 91. http://www.fao.org/3/x5989e/X5989e01.htm

Huss, H.H. 1995. Quality and Quality Changes in Fresh fish. Chapter 8. Assessment of Fish Quality. FAO Fisheries Technical Paper – 348, FAO of the UN, Rome

Martinsdóttir E, Schelvis R, Hyldig G and Sveinsdóttir K. 2009. Chapter 19. Sensory evaluation of seafood: general principles and guidelines In: Handbook of Fishery Products: Quality, Safety and Authenticity. edited by Rehbein H, & Oehlenschläger J. John Wiley & Sons Ltd, United Kingdom.

Karlsen JD, Krag LA, Albertsen CM, and Frandsen RP. 2015. From Fishing to Fish Processing: Separation of Fish from Crustaceans in the Norway Lobster-Directed Multispecies Trawl Fishery Improves Seafood Quality. PLoS ONE 10(11); e0140864.

2

Pre-harvest and Post-harvest Factors Affecting Quality of Fish

Introduction

Quality of fish is affected by various factors, which are classified into two categories: intrinsic factors and extrinsic factors. Intrinsic factors are the chemical and physical properties inherent to fish. Extrinsic factors are treatments given to fish before and after catch until it reaches the consumer, which includes the method of catch, cross contamination, processing, preservation, and product storage environment (temperature, humidity, packaging). The processing methods such as drying, smoking, chilling, and salting greatly help to minimize the spoilage. The quality of fish is also dependent on cold chain, in which the temperature is managed and monitored at every stage of production, transport, storage and sale.

2.1. Pre-Harvest Factors

Pre-harvest factors relates to the intrinsic factors affecting the quality of fish, as they are beyond the direct control of human. Various intrinsic factors affecting quality are fish species, size, sex, physiological state, parasites, naturally toxic fish, pollutants, and occasional peculiarities. Additionally, some of the extrinsic factors like method of catch and feeding interruption for starvation are also considered as pre-harvest factors affecting the quality of fish.

2.1.1. Species Characteristics

The quality of fish is related to species. For instance, flat fish keep better than round fish. Post-mortem pH varies between species. For example, long rigor period and low pH (5.4-5.6) of very large flatfish and halibut give relatively long iced storage life of 21 days, while mackerel with low post-mortem pH has no effect on shelf life. Some fish are rated as good quality as they are expensive, example, seer fish and pomfret. Fatty fish are rejected sensorial before lean fish due to the appearance of oxidative rancidity. So, lean fish keep longer than fatty fish under aerobic storage. Bony fish are edible longer than cartilaginous fish. Skin of fatty pelagic fish is very thin and contribute to faster spoilage rate, as it allows enzymes and bacteria to penetrate more rapidly. Thick skin and slime of flatfish contribute for better keepability, as slime contains bacteriolytic enzymes, antibodies and other antibacterial substances.

Quality of fish is also influenced by the size. Larger fish spoil more slowly and possess longer shelf life than smaller fish. The large surface/volume ratio makes it less prone for easy access by spoilage bacteria. Larger size fish fetch better price because of higher yield, lesser handling cost, better keeping quality and producing more uniform products, than smaller ones. Nevertheless, in some instance, larger ones are not preferred to control fill weight e.g. canned shrimps.

Sex of fish also determines the quality. Female sturgeon is more valued for their roe or caviar than male counterpart. Female freshwater eel is tastier than male eel. Female crabs are sweeter, while male crabs have better flavor. Female pink salmon tastes better than males, as males store fat in the hump. Female lobsters are considered a delicacy due to meatier tail and roe, while male lobsters have larger claws. Female octopus is tastier because of its softer meat and thinner muscle fiber. Sometimes, female octopus and snapper are not tastier right before the spawning season since most of its nutrition goes for egg development.

These quality changes are attributed to the major changes or shift in the enzymatic activities occurring during spawning. Female seabass has higher liver fructose 1,6 bisphosphatase activity than males, so females require more glucose during spawning. Phosphofructokinase activity is measured as an index of glycolytic pathway, 3-hydroxyacyl CoA dehydrogenase for □ oxidation, citrate synthase for citric acid cycle, and cytochrome oxidase for oxidative phosphorylation. At the time of spawning, 50% reduction in glycolytic capacity and a 4 folds increase in oxidative capacity are observed. The brain cytochrome P450 aromatase activity undergoes 50-fold activity at gonadal growth in goldfish; and lipoxygenase activity is greater in salmon during spawning.

2.1.2. Composition and Physiological State

Composition and physiological state affect the quality of fish. Fish after spawning have higher water content and lesser fat content. Sardines and mackerel have very high fat content during winter months. Crabs contain high meat content during new moon period. Fish with more glycogen turn acidic due to the production of lactic acid.

Heavily fed fish, if un-gutted leads to a problem, known as **"belly burst"**, while held in ice. It occurs due to the digestive proteolytic enzymes present in the gut. The extensive autolysis caused by enzymes results in production of low molecular weight peptides and free amino acids, which helps in proliferation of spoilage bacteria and production of gases such as CO_2 and H_2. The muscle softening subsequently occurs and ruptures the belly wall due to low post-mortem pH, and allows the blood to drain out. Belly bursting occurs more common in pelagic fish species such as herring, sardine and capelin. It is more predominant in summer months, in fish feeding on "red feed" containing copepods and euphausiids. In herring and sardine, autolysis of gut cavity occurs due to physical handling practices such as freezing/ thawing, and time of ices storage.

"**Gaping**" phenomenon occurs in fish due to the action of collagenases. Fish muscle cell block or myotome is surrounded by connective tissue (collagen fibrils) called myocommata. Chilling of fish deteriorates collagen fibrils in fish muscle by solubilizing type V collagen by the action of autolytic collagenase enzymes. The breakdown of myotome causes "gaping" during long-term storage in ice or short-term storage at high temperature. It leads to shortening of the muscle post rigor.

"**Chalkiness**" is a special problem that occurs due to more acidic pH. It occurs more common in seer fish and halibut. It turns the flesh into white opaque colour and tough during storage. It is more common in summer months. Chalkiness occurs due to the buildup of lactic acid and low post-mortem pH. It causes denaturation of muscle proteins causing extensive drip loss and turns the muscle opaque. It occurs seven days after the death of fish. Male fish tend to be chalkier than female fish.

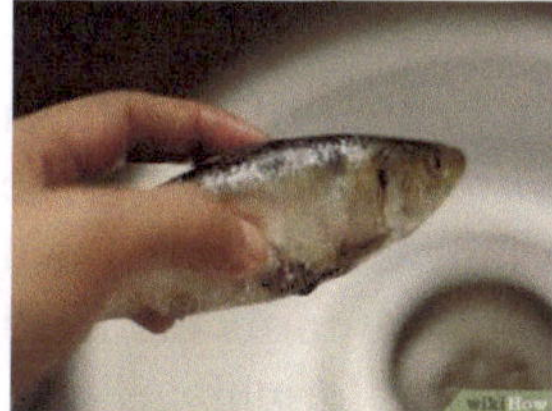

Belly bursting in sardine

Gaping in fish fillet

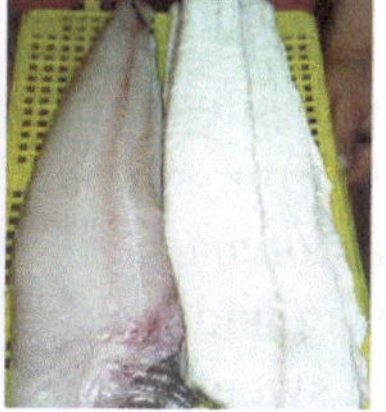

Chalkiness in fish fillet

2.1.3. Parasites and Other Organisms

Incidence of parasites in fish is an important quality factor. Parasites are mainly located in the head and viscera of fish. Parasites in fish are mainly protozoa, flat-worms (platyhelminths), round worms (nematodes), flukes or tape worms (cestodes) and certain crustacea. Occurrence of parasites in fish is a natural phenomenon. However, thorough cooking kills all parasites and render them harmless. Health problems occur only raw fish or semi-preserved fish such as cold smoked products or sashimi are consumed.

The most common parasites present in marine food fish are nematodes. They are referred as "herring worms" or "cod worms." All of them belong to the family Anisakidae. *Anisakis* spp. is common in squids and herrings. Cod worms (*Pseudoterrnova decipiens*) are present either in gut or viscera or encysted in the flesh. Freshwater and migratory fish carry small white flabby tapeworms (*Diphyllobothrium* spp.), which are common in salmon. Fish gets infected when they eat small crustaceans carrying parasite eggs or larvae. The worms penetrate to the muscle or internal organs and develop into an adult to infect human. Older fish tend to have more infestation than young fish.

The consumption of nematodes causes severe gastric upset called "Anisakiasis". They cause severe abdominal pain and intestinal upset that can last for 10 days. The tapeworms, on the other hand can live in the intestinal tract for several years. The infection is not fatal, but causes symptoms like abdominal pain, weakness, loss of weight and anemia.

Cooking the fish to an internal temperature of 60ºC is sufficient to kill all nematodes and tapeworms. Hot smoking done at 71ºC is also sufficient to kill parasites. Hard-salting (curing) fish kills parasites, but pickling without salt can never destroy parasites. Consumption of raw fish is traditional in some countries. In such a case, freezing at freezer temperature of -20ºC for 24 h kills all parasites. Large fish require 5 to 7 days of freezing. Some processors remove parasites in the gut and belly walls during the filleting process. Fish fillets are also examined by **candling process**, but fish larvae present 1 cm deep is difficult to detect. The consumers are advised to cook all fish until it flakes and loses its translucent texture. The consumers are raw fish or cold smoked fish are advised to use only frozen marine fish.

Other parasites are myxosporea infection that cause softening of fish flesh (eg. *Chloromyscium thyrsites),* lung fluke (*Paragonimus* sp.), and sea louse (*Lepeophtheirus* sp.) found in fresh salmon and trout. A fungal disease, known as 'greasy haddock' is caused by *Ichthyosporidium hoferi* that turns the flesh soft and sweetish with slight sickly smell. Fish infected with *Aeromonas* spp. has lesions, nodules (small lump or swelling) and pustula (pimple or blister) areas in the skin and flesh.

2.1.4. Naturally Toxic Fish

Some fish are naturally toxic and cause injury to health or even death. Ciguatera fish poisoning (CFP) occurs by consumption of certain carnivorous fish (barracuda, snappers) harvested from shallow waters in or near tropical and subtropical coral reefs in certain season. Ciguatoxin (CTX) is produced by a benthic dinoflagellate, *Gambierdiscus toxicus* is responsible. Normal cooking does not destroy the toxin.

Puffer fish poisoning (PFP) occurs by consuming the puffer fish flesh contaminated with viscera. Tetradotoxin (TTX) is produced by bacterium, *Vibrio alginolyticus*. The mortality rate is 50% and the largest incidence is reported from Japan.

Paralytic shellfish poisoning (PSP) occurs by eating certain molluscs, particularly mussels and clams. The toxin is called as mussel poison or mytilo toxin / clam poison or saxitoxin. The toxin is produced by dinoflagellates, such as *Gymnodium* spp. or *Ptychodiscus* spp. When these organisms dominate, the water assumes a reddish tinge, causing the "**red tide phenomenon**". The protective measure is not to collect shellfishes from such waters. The other sources of PSP are *Gonyaulax catenella, Alexandrium* sp. and *Pyrodinium* sp.

Diarrheic shellfish poisoning (DSP) is caused by toxins, okadic acid (OA), and dinophysis toxin; and the sources of DSP is *Dinophysis* and *Prorocentrum*. Neurotoxic shellfish poisoning (NSP) is caused by *Gymnodium breve* or *Ptychodiscus breve* and the toxin responsible is brevetoxin.

Venerupin shellfish poisoning (VSP) is also known as oyster or asari poisoning and the source is *Prorocentrum* sp. Amnesic shellfish poisoning (ASP) is caused by the toxin, domoic acid that affects the brain leading to memory loss. The toxin is produced by a diatom. Erythematous shellfish poisoning (ESP) affects the circulatory system by attacking the red blood cells.

2.1.5. Contamination with Pollutants

Contamination of fish with pollutants is another quality problem difficult to identify by the consumers. The main classes of pollutants are metals, chlorinated hydrocarbons, mineral oils, radioactive isotopes and micro-organisms. Fish harvested from polluted waters contain a large number of harmful heavy metals such as mercury, cadmium, lead, selenium and arsenic. The important persistent organic chemicals present in fish are a group of chlorinated hydrocarbons including insecticides/pesticides such as DDT, aldrin, dieldrin, benzene hexachloride (BHC or lindane) and polychlorinated biphenyls (PCB's). The maximum permitted levels for DDT and dieldrin in most fish as 5.0 and 0.30 mg per kg wet weight, respectively. The residues of antibiotics such as chloramphenicol, nitrofurans, tetracyclines, etc. are present in the flesh of cultured fish and shellfish and they are dangerous to humans. Occasionally, fish catches are grossly contaminated with chemicals, especially mineral oils resulting from accidental (during offshore oil exploration and oil tanker movements) or other kinds of large-scale release, which causes tainted odour or flavour in fish. Fish exposed to high energy radiation originating from radioactive isotopes are extremely injurious to health. Naturally occurring radioactive isotopes occur at very low concentration in fish. In some areas, artificial isotopes are released in very small amounts into the environment.

The influence of environmental contaminants alters the enzyme content of fish tissues. The presence of specific enzyme isoforms is an indication of environmental pollution in fishing grounds. Fish subjected to oxidative stress due to contaminants such as polyaromatic hydrocarbons, polychlorinated biphenyl, and pesticides, show an increase in antioxidant enzymes. Glutathione peroxidase, superoxide dismutase, catalase, glucose-6-phosphate dehydrogenase, and glutathione reductase are useful indices. Specific Cu and Zn-superoxide dismutase isoforms are biomarkers of oxidative stress caused by transition metals or organic xenobiotics. A decrease in peroxidase activity noticed in murrel inhabiting polluted waters. An increase in acid phosphatase activity is a general marker for sub-lethal cadmium concentration in mollusks, crustaceans and fish. Gill and serum acid phosphatase increased in carp exposed to vegetable oil factory effluent, but not liver; and hence a suitable index of copper stress in carp.

Fish exposed to anti-fouling compound, triphenyltin showed reduction in chymotrypsinogen and trypsinogen content. Endosulfan, an organochlorine pesticide reduced the activity of cytoplasmic and mitochondrial malate dehydrogenase in catfish, by forming an inhibitor complex with the enzyme. Long term exposure of rainbow trout to hexachlorobiphenyl induces liver cytochrome activities (arylhydrocarbon hydroxylase and ethoxyresorufin-O-deethylase). Ethoxyresorufin-O-deethylase is the most sensitive and rapid indicator of dioxin and metal pollution in carp and trout.

2.1.6 Contamination with Pathogenic Microorganisms

Fish harvested from polluted waters having raw or inadequately treated sewage introduces pathogenic organisms injurious to health. Sewage contains bacteria and virus of fecal origin. *Salmonella* causes typhoid and paratyphoid, while *Shigella* cause dysentery. The only infectious virus is hepatitis A, which transfer to fish, crustaceans and molluscans from near sewage outfalls or sewage polluted river, via. seawater and detritus. The molluscans harvested from the sewage-polluted areas cause health hazard, if they are consumed raw. The indicator organisms determined to assess the quality are fecal coliforms or *Escherichia coli.*

2.1.7. Occasional Peculiarities

Some occasional peculiarities like tumors, ulcers, nodules, abnormal colouration, abnormal odours, etc. rarely observed in fish also affect the quality. Physical damage occurs in food fish by predators such as seal on salmon, and shark on many species before capture, but do not pose a serious problem in processing. The larvae of *Mytilus* spp causes a bitter taste in herring. The flesh of cod and other gadoids occur in various shades of pink, due to the presence of astaxanthin and zeaxanthin derived from feed or some metabolic disturbance. The pigment turns yellow on oxidation.

Fish caught in certain regions have off-flavours. Off-flavours are attributed to their feeding. Cod and other gadoids feed on planktonic mollusks, pteropods (*Spiratella helicina)*, which contain dimethyl - β-propiothetin. This is converted to dimethylsulphide (DMS) in fish causing peculiar odours and flavours like 'black berry, weedy, petrol, diesel, mineral oil, iodine and sulphide'.

Freshwater fish, tilapia form an earthy or muddy odour and flavour, which affects consumer acceptance. Muddy-earthy taint off-flavour occur is caused by two compounds: geosmin (trans, 1-10 □ dimethyl 9 □-decalol) and 2-methyl isoborneol, MIB, which originates in waters having high concentration of actinomyces, *Streptomyces* and *Actinomyces*. High concentration of geosmin causes an earthy-musty flavor in the tail muscle of the penaeid shrimp grown in coastal culture ponds.

An iodine-like flavour is found in some fish and shrimp, which is caused by volatile bromophenolic compounds produced by marine algae, sponges, and Bryozoa. Oil taint is found in fish caught in areas where large oil spills occur. The fraction of crude oil soluble in water gives the off-flavour due to accumulation of hydrocarbon compounds.

2.1.8. Fish Catching Methods

Wild fish are harvested by various kinds of nets, hooks, and traps. The wild fish undergoes struggle at the time of capture followed by a period of asphyxiation on board. The catch quality of gillnet caught fish is high. However, gillnets operated with several days soak time produce inferior quality fish catch, as fish caught early may die and start to deteriorate long before the nets are retrieved.

The catch quality of pelagic trawl is medium, because majority of the fish catch undergoes stress fighting for life and the damage to fish catch caused by physical pressure at the bottom of the trawl. Such fish when held it ice, the quality deteriorates faster. Tuna caught in a highly stressed condition and wild salmon caught by gill netting die after stress exhaustion produce lactic acid in the muscle at high temperature and develop a dull muscle, acidic and metallic taste.

Longline caught fish are in general of high quality, but long soak times will lead to inferior fish quality. Pot fishing has high energy efficiency and the catch quality is good to superior, as the catch normally remains alive and in good condition. The catch quality of trap fishing and fishing with spear or harpoon are high. Selection of fishing method based on the catch quality is therefore more appropriate.

The stress fish undergoes induce the activity of certain anti-oxidative enzymes in the muscle that influence lipid peroxidation after harvest. Glycogen phosphorylase and glycogen phosphorylase kinase are the two key glycolytic enzymes in fish for its movement. During exhaustive stress, the full cascade of glycogen phosphorylase (GPK, cAMP dependent GK, and cAMP levels) gets activated by Ca^{2+} and cellular metabolites in fish muscle and liver. Exhaustive stress has also modified rainbow trout muscle AMP deaminase by a reversible phosphorylation, which converts AMP to IMP. Presence of high IMP has got a direct effect on the consumer acceptable of seafood.

2.1.9. Starvation

In heavily fed aquaculture fish, the digestive tract contains many bacteria that produce digestive enzymes and leads to intense postmortem autolysis. The enzymatic activity leads to development of strong odors and flavors in the abdominal area. Starvation is important to prevent feces trailing from the anus, as it is objectionable for consumers. So, starvation period of 1 to 3 days is recommended depending on temperature. For instance, 1 day is the minimum feeding interruption period in sea bream, and 8 d is the maximum. Feeding interruption is not advisable beyond 2 d, as it affects the market price for the fish. Starvation can result in 1% of weight reduction at higher temperature (35°C) than at 20°C. The muscle texture of sea bream turns firmer when starved for 8 d, as compared to 1 to 3 d, due to changes in protein solubility and pH.

Starvation influences the quality of fish in many ways. Starved fish contain reduced muscle glycogen, so it alters pH decline and rigor onset in fish. Some fish is less susceptible to gaping eg. cod. Starvation increases □-amylase activity in muscle of catfish (*Heteropneustes fossilis*), shifting to glycogenolysis pathway. Re-feeding fish reduced the tissue amylase activity and shifted to phosphorylase pathway. Starvation also reduces hepatic enzymatic activity (glucose-6-phosphate dehydrogenase, 6-phosphogluconate dehydrogenase, malate dehydrogenase, glutamate dehydrogenase and alanine aminotransferase involved in amino acid metabolism) in rainbow trout, causing hepatic protein repression. Starvation reduced soluble NADP isocitrate dehydrogenase involved in lipogenesis in carp and tilapia. Prolonged starvation also resulted in the decrease of NAD-isocitrate dehydrogenase involved in oxidative phosphorylation.

2.1.10. Harvesting of Fish

Harvesting is one of the most important activities of fish farming. It is an important to get fish in good condition and quality. The key causes for food loss and waste linked to quality and the mortality of fish during live fish marketing are:

- Improper starvation before harvesting
- High stocking density in sample holding tanks
- Slow and stressful harvesting
- Contamination with mud and dirt from pond environment
- Contact with unclean surfaces and unclean staff
- Improper washing before icing
- Delay in chilling
- Physical damage during harvesting and handling (bruising and damage to fish flesh)
- Delay in bleeding or inadequate bleeding
- Malfunction of live fish holding and transport technology

2.2. Post-Harvest Factors

The post-harvest extrinsic factors that affect the quality of fish are the temperature, pre-processing conditions, processing ingredients and additives, processing methods, and packaging methods. Cleanliness and sanitary conditions also affect the quality of fish products that are handled and processed. However, the post-harvest factors can be controlled to maintain the quality of fish.

2.2.1. Killing Method

Stress undergone by wild or farmed fish before slaughter affect the quality of the fish. The degree of muscle activity before slaughter affects the firmness of fish flesh during rigor mortis. When the fish is killed after the muscle activity at the time of capture, the muscle cells produce more lactic acid due to anaerobic respiration. The synthesis of ATP stops and rigor mortis sets in faster.

Spiking of tuna

Spiking (iki jimi in Japanese) is a method used to kill the fish onboard the vessel by puncturing the brain, which prevents the muscle activity and delays the onset of rigor mortis as compared to **hypothermia** i.e., immersion of fish into chilled water for slower death. As a result, the loss of ATP is stopped and interlocking of thin and thick filaments is prevented to form rigor mortis.

When the fish is killed by spinal cord destruction as comparted to struggling or temperature shock, the lactic acid production is low and onset of rigor mortis is slowed down.

In highly stressed fish, all muscles enter rigor very quickly and the whole fish becomes very stiff and turns difficult to process further. In fish with a low level of enzymatic activity, only few muscles enter rigor mortis, while the others do so later. If difference occurs in enzymatic activity, not all muscles are in rigor at the same time and the whole fish is less stiff. Pre-mortem handling stress also affects color of salmonids flesh. The loss of color is caused by solubilization of muscle proteins due to low pH, leading to drip loss in the pre-rigor fish.

2.2.2. Temperature

Temperature is the most important single factor affecting the degree of freshness of fish. Fish spoilage starts immediately after death by a combination of microbial, chemical and autolytic reactions. All these processes need optimal temperature range for their action. If the temperature is within the optimal level, the fish lose its freshness faster. The exposure of fish to high temperature increases the rate of bacterial growth, enzymatic activity, and oxidation leading to rapid spoilage to reduce their shelf-life and cause food safety risks. The storage temperature can reduce the effects produced by pre-slaughter stress.

The fish quality is controlled by reducing the temperature to keep them fresh. As the temperature decreased, there is a reduction in ATPase activity. The fish take longer time to spoil at low temperature. It is a common practice to store fresh fish in ice (0°C), however temperature requirements vary among food items. Refrigerated foods are held in freezer storage temperature at –18°C or below; chilled temperature at 0°C to 1°C; medium-chilled at 5°C; and exotic-chilled at 10°C to 15°C.

In the temperature range of 0 to 25°C, the microbial activity is relatively more important. Many bacteria are unable to grow below 10°C, even psychrotrophic bacteria grow very slowly. At 0°C, the growth of fish spoilage bacterium, *Shewanella putrefaciens* is less than one-tenth of the rate at optimum temperature.

Skeletal muscle undergoes modification in mitochondrial content and oxidative metabolism at low temperature acclimation. In cold acclimatized fish, increase in enzymatic activity and changes in the kinetic properties of enzymes takes place. Mitochondrial enzyme markers are high in red and white muscle of cold acclimated fish. There is an increase in cold adaptive enzymes. The rate of muscle ATP catabolism is greater in cold water fish than in warm water fish. Hepatopancreatic enzymes, particularly lipolytic enzymes such as glucose 6 phosphate dehydrogenase and NDPH malate dehydrogenase are also abundant in cold acclimated fish. So, the activity of cold isozyme of muscle lactate dehydrogenase in higher in fish. Ca^{2+} ATPase activity is greater in cold acclimated trout, but not in carp. Lactate dehydrogenase increased in skeletal muscle of loach at low temperature (5°C) through adaptation, while acclimation of tilapia to low temperature (20°C) decreased muscle LDH activity.

2.2.2.1. Fish Spoilage Models

Mathematical RRS models are developed to predict the shelf life of fish at different storage temperatures. The types of chemical reactions that cause spoilage are not taken into account in these models. In order to predict shelf-life at different temperatures one has to provide the shelf-life of the product at a single storage temperature. The RRS models then predict shelf-life at different temperatures. For fresh fish, RSS at t°C is defined as the shelf-life at 0°C divided by the shelf-life at t°C. In this case, 0°C is taken as a reference temperature for fresh fish. In case of lightly preserved foods, 5°C is used as reference temperature.

$$\text{Relative rate of spoilage at T°C} = \frac{\text{Keeping time at } 0^{\circ}\text{C}}{\text{Keeping time at } t^{\circ}\text{C}}$$

The effect of temperature on RRS is more or less similar for fresh fish, although broad differences are observed in shelf lives of various fish products (Table 2.1).

	0°C		5°C		10°C	
	Shelf life (d)	**RRS**	**Shelf life (d)**	**RRS**	**Shelf life (d)**	**RRS**
Crab claw	10.1	1	5.5	1.8	2.6	3.9
Salmon	11.8	1	8.0	1.5	3.0	3.9
Sea bream	32.0	1	-	-	8.0	4.0
Packed cod	14	1	6.0	2.3	3.0	4.7

The effect of temperature on the rate of chemical reactions is described by the Arrhenius equation. The effect of sub-optimal temperature on growth of microorganisms is studied by 2-parameter square root model.

$$\sqrt{1/_{max}} = b(T - T_{min})_{6b}$$

Where, T is the absolute temperature (°K) and T_{min} is theoretical minimum temperature of growth. The effect of time/temperature condition on product shelf life is cumulative. The spoilage models help to predict the remaining shelf life of the product. Electronic time/temperature integrators are developed to predict RRS accurately. The predicted shelf lives of fish products stored at different temperatures (Table 2.2).

Shelf life in days of product stored in ice (0°C)	Shelf life at chill temperatures (days)		
	5°C	10°C	15°C
6	2.7	1.5	1
10	4.4	2.5	1.6
14	6.2	3.5	2.2
18	8	4.5	2.9

The temperature history of a product through a distribution system is determined by a temperature logger. A product temperature profile allows the estimation of growth of pathogenic microorganisms to predict safety models.

The microflora responsible for spoilage of fresh fish changes with the storage temperature. At low temperatures (0-5°C), *Shewanella putrefaciens, Photobacterium phosphoreum, Aeromonas* spp. and *Pseudomonas* spp. cause spoilage. At high temperatures (15-30°C), different species of Vibrionaceae, Enterobacteriaceae and Gram-positive bacteria cause spoilage. First, the estimates of RRS for whole fresh fish, for packed fresh fish and superchilled fresh fish products are obtained. The average RRS of a large number of species stored at 20-30°C are 25 times higher than at 0°C. The RRS of tropical fish is more than twice as high as those estimated from the temperature models of temperate fish. A tropical spoilage model covering the temperature range from 0-30°C was developed by Dalgaard and Huss (1994). The natural logarithm of RRS of tropical fish linearly relate to the storage temperature. Temperature models do not consider the initial product quality. So, the shelf life is determined at a given temperature at a given initial quality by the following equation.

$$\text{Shelf life} = \frac{\text{Final - Initial level of a quality indicator}}{\text{Rate of spoilage at the actual storage condition}}$$

Predictive microbiology is a pro-active approach to fish quality and safety, as it provides information on bacterial responses related to environmental factors through mathematical models. Mathematical models for growth, survival or inactivation of microorganisms are valuable tools to assess safety and shelf-life of food. Models are available to predict food safety including those that predict histamine formation in marine finfish.

2.2.2.2. Delay in Chilling

The delay before chilling affects the quality of fish. A white-flesh lean fish when enter rigor mortis at temperatures +17°C, their muscle tissue ruptures severely through muscle contractions, the connective tissue weakens and flakes of fillets separates from each other causing "**gaping**" phenomenon. Such fish are difficult to fillet. Rapid chilling is also crucial for fatty fish to maintain quality. Herring and garfish exposed to sun and wind for 4-6 hours before chilling had reduced storage life, due to oxidation of lipids and development of rancid off-flavours. The high temperature accelerates the oxidation process in fish.

2.2.2.3. Super Chilling

Super chilling of fish between 0°C and -4°C extends the shelf life of fish products. The square root spoilage model gives the RRS of super chilled products. The shelf life predicted at -1°C, -2°C, and -3°C for a product that keeps 14 days in ice is 17, 22, and 29 days, respectively. Super chilling is used in areas where productive fishing grounds are far from ports and consumers. Super chilling is also used in transport of live fish in Japan. The negative effects of super chilled fish are drip loss, loss of appearance and texture of cod and haddock. Appearance and texture of cod stored at -2°C for 10 days is inferior to those stored at 0°C in ice. At -3°C, the texture of fish turns unsuitable for filleting. The drip loss occurs due to formation of large ice crystals, protein denaturation and increased enzymatic activity in the super chilled or partially frozen fish. Super chilling is good for few species such as tuna and salmon.

Refrigerated sea water (RSW) used in fishing vessels has sub-zero temperature obtained by reducing the freezing point of water by NaCl or other freezing point depressors. RSW chill fish more rapidly by reducing the exposure to oxygen, reducing the pressures on fish by ice and saving labour cost. RSW chilled fish gives a salty taste due to take up of seawater.

2.2.3. Hygienic Handling

Hygienic handling of fish from the time of catch ensure good quality and long storage life. Rough handling and bruising, results in faster spoilage, as it allows contamination of fish with bacteria and allow releases of enzymes. Careless handling can also burst the guts and spreads the content into the flesh.

The use of clean fish boxes minimizes bacterial contamination. While fish iced in boxes are stacked on top of each other, there must be provision to drain the ice melt-water into the box underneath it to avoid bacterial contamination. Melt water contains a large number of bacteria. The use of radioactive irradiation at 100-200 krad helps to reduce the number of bacteria and prolong storage life. The process is costly and is unacceptable by many people due to fear of health concern. The incorporation of antibiotics in ice to preserve fish are also not acceptable due to public health concern. The recent method is use of CO_2 treatment either in containers with chilled seawater or as part of a modified atmosphere during distribution or in retail packages. The washing with chlorinated water decontaminates fish. The amount of chlorine creates off-flavours in fish meat. The newly caught fish should be washed in clean seawater without any additives. The purpose of washing is to remove visible blood and dirt.

The use of functional ice (FICE) is the latest innovation over regular ice. It is made by freezing water with antimicrobial substances. FICE slowly releases the antimicrobial over the good surface and eliminates the spoilage microorganisms throughout the storage. FICE prepared by freezing an aqueous solution of sodium tripolyphosphate (5%) and sodium lactate-sodium diacetate (1% and 2.5%) reduced *Salmonella* Typhimurium counts more than 1 log CFU in poultry meat.

2.2.4. Pre-processing

The quality of fish decrease, if gutting is done prior to processing. During heavy feeding periods, many bacteria are present in the digestive system and strong digestive enzymes are produced by the fish. The main reason for gutting is to prevent autolytic spoilage caused by enzymes rather than bacterial spoilage. The enzymes lead to autolysis, giving rise to strong off-flavour in the belly region, and finally leading to belly burst. Gutting also exposes the belly area and cut surfaces to the air, thereby making them more susceptible for oxidation and discoloration.

Factors such as species, age, amount of lipid, catching method and ground, should be taken into consideration to decide on gutting. Small and medium sized fatty fish such as herring, sardines, and mackerel are not gutted immediately after catch, as they are small fish and cause problems like discoloration and rancidity. But, heavily fed fish, if ungutted leads to a problem, known as "belly burst". Belly burst occurs in a well-fed fish, when the strength of connective tissue decreases at post-mortem low pH. Gutting is done in lean species to avoid the quality loss and reduction of shelf life of 5 or 6 days. In lean fish, after two days from catch, discoloration occurs in belly area and raw fillet acquires an offensive cabbage odour.

Gutting is done traditionally done by hand with a knife but only in large vessels, machines are used. Industrial gutting and beheading are mechanized in developed countries on board. The machines perform gutting by sucking the viscera out and cleaning the belly part through the mouth. Fish should be thoroughly washed to remove traces of blood and debris and to wash bacteria and intestinal content out of the gut cavity, skin, and gills of the fish, once gutting is performed.

Slaughtering by electrical stunning produces active movement to break open vertebrae and ruptures the blood vessels and forms blood spots. Electrical stunning is a better method for salmon than CO_2, as it causes an earlier onset and resolution of rigor mortis. Bleeding is frequently done in large farmed fish. It is therefore recommended to cut the gills with a sharp knife, which allows the fish to swim and die due to anoxia caused by blood loss.

2.2.5. Packaging Conditions

Packaging is more convenient to handle fish products in supermarkets. Packaging maintains product quality, offers protection, facilitates handling and movement of goods. Fish processors and fish exporters prefer bulk packaging that is robust enough to withstand the demands of world seafood trade. For the retail trade, a wide range of packaging are available - vacuum packaging (VP), modified atmospheric packaging (MAP), skin packaging, sealed plastic trays, folding cartons and tins. Fresh fish is packed differently from frozen fish, smoked or marinated fish, or ready-to eat (RTE) fish. Some products are packed several times. For instance, MAP smoked salmon is additionally packed in a cardboard case. According to EEC Council Directive 91/493/EEC of 22 July 1991, vacuum packaged (VP) and modified atmospheric packaging (MAP) fish products are used for fresh products. Packaging controls both microbial and non-microbial spoilage in fish.

2.2.5.1. Microbial Spoilage

Carbon dioxide is used for preservation of fresh fish products to extend the shelf life. High CO_2 concentration can reduce microbial growth and extend the shell life of fish. Modified atmospheric packaging (MAP) with high CO_2 levels (25% to 100%) increases the shelf life of fish by few days, while vacuum packaging (VP) does not increase it. The difference in spoilage flora is the reason for the differences. Spoilage of fish under aerobic conditions in ice is caused by gram negative bacteria, like *Pseudomonas* spp., while in VP and MAP products, gram-positive bacteria (mainly lactic acid bacteria) dominates. In VP cod, the gram-negative bacteria, *Photobacterium phosphoreum* is identified as specific spoilage organism (SSO), as it is highly resistant to CO_2. *P. phosphoreum* reduces TMAO to TMA and produce little H_2S in fish substrates. The bacterial population remains very low (10^5-10^6 cfu/g) at the time of sensory rejection of MAP fish products. It indicates that non-microbial reactions are responsible for spoilage of VP or MAP products.

2.2.5.2. Non-microbial Spoilage

The CO_2 dissolves in the water phase of MAP fish and decreases the pH by 0.2 to 0.3 units. The water holding capacity (WHC) of muscle proteins decreases at decreased pH and an increased drip loss occurs in cod fillets, shrimps, and salmon, but not in herring, red snapper, trevally, crab and rockfish. The texture quality of fish held in 100% CO_2 get reduced, however no negative effect observed up to 60% CO_2. The colour of the belly flaps, of cornea and of the skin altered for whole fish stored in high CO_2. MAP stimulates the formation of metmyoglobulin in red fleshed fish and darken the fish muscle. If oxygen containing MAP is used, the development of rancid off-odours in fatty acid does not occur.

The CO_2 in combination of refrigerated seawater (RSW) is used to extend the shelf life of some fishes such as Pacific cod, herring, rock fish, shrimp and chum salmon. The CO_2 acidifies the seawater, lowers the pH, and inhibits the enzymatic reactions causing black spots in shrimps. The negative effects are unacceptable tough texture and "soft shell" appearance in shrimps, and also it dissolves flesh of mackerel by making it unsuitable for canning.

References

Dalgaard, P. and Huss, H.H. 1994. Mathematical modelling used for evaluation and prediction of microbial fish spoilage. In: D.E. Kramer, F. Shahidi and Y. Jones (eds.) Proceedings of the Symposium New Developments in Seafood Science and Technology, CIFST, Vancouver, Canada.

Haard N.F. and Simpson B.K. 2000. Seafood Enzymes: Utilization and Influence on Post Harvest Seafood Quality. CRC Press. P. 696.

Huss H.H. 1995. Quality Changes and Shelf Life of Chilled Fish Chapter 6. In: Quality and Quality Changes in Fresh Fish, FAO Fisheries Technical Paper-348Food and Agriculture Organization of the United Nations, Rome

Jeyasekaran G. and Jeyashakila R. 2020. Quality of Aquatic Food. Chapter 2. In: Quality Assurance of Aquatic Foods. Pp 4-15, Narendra Publishing House, Delhi (India)

Teklemariam A.D., Tessema F., and Abayneh T. 2015. Review on Evaluation of Safety of Fish And Fish Products. International Journal of Fisheries and Aquatic Sciences, 3(2) : 111-117

Thant, N. 2019. Improvement of Post-Harvest Handling of Aquaculture Fish in Myanmar. United Nations University Fisheries Training Programme, Iceland Final Project. Http:// Www.Unuftp.Is/Static/Fellows/Document/Nyo/18prf.Pdf

3

Quality Changes During Fish Processing

Introduction

Seafood undergoes rapid spoilage after capture or harvest due to their highly perishable nature. Spoilage refers to any change that renders seafood unacceptable for human consumption. Spoilage occurs as a consequence of various microbial, biochemical and chemical breakdown processes. Spoilage leads to quality changes in seafood. Quality change refers to any change that occur after harvest due to unpredictable catch, uncontrolled harvest, inadequate handling, processing and storage facilities. The initial quality loss is mainly due to the post-mortem autolytic activity (enzymatic process) and chemical degradation process (lipid oxidation); and later by microbial degradation. The quality problems occur as a result of poor-quality raw material, ice and water, microbiological contamination, presence of heavy metals and other storage aspects during processing.

Processing of fish means preparing raw fish into a marketable form. Pre-processing is done prior to final processing. Pre-processing is often done on board fishing vessel or in a shore based plant. It includes operation such as washing, sorting, grading, butchering, etc of the harvested fish. The butchering involves removal of non-edible portions such as viscera, head, tails and fins from fish. The yield of fish meat varies from 30 to 70% depending on the butchering process. Butchering process includes beheading, gutting, filleting, peeling, deveining, shucking, etc. The discarded portion of fish meat are fish wastes, often used for animal feed or for conversion to bioactive functional foods. Final processing of fish includes processes such as heating, freezing, drying, salting, marinading, irradiation, or addition of chemicals. The processed seafood is categorized into five major groups – chilled, frozen, canned, cured and other value-added products in this chapter.

3.1. Quality Changes in Chilled Fish

Chilled fish are raw fish preserved in ice (0°C) to prevent rapid spoilage. A reduction of storage temperature by 5–6°C reduces the rate of biochemical reactions and doubles the shelf- life of fish. Chilling of fish is done from landing centers to wholesalers or processing industries, and distribution to retailers. Fish frozen onboard the sea are also treated as chilled fish after thawing. The quality and spoilage of chilled fish are more or less similar to that of fresh fish.

Fish in landing centers or retailers or fish processing plants undergo further handling and exposure to high temperature. The pre-processing steps applied are gutting, filleting, peeling or shucking. A fish cut into slices, steaks or fillets are more susceptible to spoilage because of the greater surface area. The chilling process slows down the spoilage. It is difficult to bulk store fish in chilled liquid media. So, refrigerated chill room maintained at 2-3°C in conjunction with ice is used for bulk storage of fish.

Shelf- life of iced fish caught in warm, tropical waters is better than from those caught in cold, temperate waters. Fish caught in tropical marine waters keep for 12–35 days in ice, while those caught from temperate waters that keep for 2-21 days. Freshwater fish have longer shelf lives in ice than marine species. Changes in protein and lipid fraction, degradation of nucleotides, formation of volatile amines, biogenic amines, hypoxanthine (Hx) and action of certain bacteria takes place during chilled storage of fish. Fatty fish undergo spoilage mainly due to lipid oxidation.

3.1.1. Chemical Changes

3.1.1.1 Lipid Degradation

Lipid oxidation is a dominant spoilage process in chilled fatty fish such as trout, sardine, herring and mackerel, as both hydrolytic and oxidative rancidity occur. In advanced stages of oxidative rancidity, hydroperoxides formed breakdown and generate low molecular-weight carbonyl and alcohol compounds to give objectionable rancid odors and flavors. Prevention is by the use of slurry ice.

3.1.1.2. Volatile Bases Formation

Many spoilage bacteria utilize trimethylamine oxide (TMAO) under anaerobic condition to produce trimethyl amine (TMA) by the action of TMA-oxidase. In later stages of storage, odorous low-molecular-weight sulfur compounds such as H_2S, CH_3SH (methyl mercaptans), volatile fatty acids and ammonia are produced. Delay in icing and exposure to ambient temperatures encourage proliferation of spoilage bacteria.

3.1.1.3. Biogenic Amines Formation

Scombroid fish contain high levels of free histidine, which gets converted to histamine by the action of histidine decarboxylase produced by *Vibrio*, *Proteus morganii* and *Klebsiella pneumoniae* at temperatures between 8°C and 15°C.

Other biogenic amines are cadaverine from lysine, putrescine from ornithine, and tyramine from tyrosine. Psychrotolerant bacteria viz. *Photobacterium phosphoreum* present in fish produce histamine at 2°C.

3.1.2. Microbial Changes

Spoilage of chilled fish is mainly by the dominant spoilage psychrotrophic bacteria viz. *Pseudomonas* spp. and *Shewanella putrefaciens*. *Shewanella* spoilage is characterized by TMA and H_2S, whereas *Pseudomonas* spoilage is characterized by sweet, rotten sulfhydryl odors. The psychrophilic sulfur-producing bacteria, *Pseudomonas* and *Aeromonas* spp. occur in chilled freshwater fish.

3.1.3. Enzymatic Changes

3.1.3.1. Nucleotide Degradation

The initial quality loss in chilled fish is due to degradation of nucleotides (ATP-related compounds) by autolytic enzymes. Adenosine triphosphate (ATP) degradation proceeds with the formation of intermediate products such as adenosine diphosphate (ADP), adenosine monophosphate (AMP), inosine monophosphate (IMP), inosine (INO), and hypoxanthine (Hx) by autolytic enzymes. The absence of IMP is responsible for the loss of fresh fish flavor.

The muscle ATPase activity is controlled through the modulation of calcium (Ca^{2+}) ions. The muscle ATPase is present in globular head of myosin molecules (heavy meromyosin) and active in high levels of calcium. The Ca^{2+} activated ATPase breakdown ATP to release energy during contractile process. Myokinase is a potential inhibitor of ATPase. So, myokinase reaction produces AMP during muscular activity from ADP and ATP. Then, AMP deaminase acts on AMP to produce IMP and ammonia. The IMP is converted into inosine (Ino) and hypoxanthine (Hx), where ammonia is transported to gills through blood for excretion. Three enzymes are involved for the conversion of IMP to Ino such as 5'nucleotidase, alkaline phosphatase and acid phosphatase; and the former is the most important in postmortem fish. Two enzymes, namely nucleoside phosphorylase (NP) and nucleoside hydrolase or inosine nucleosidase (IN) catalyzes degradation of Ino to Hx. NP catalyzes the reaction of Ino to Hx and ribose 1 phosphate, while IN catalyzes the reaction of Ino to Hx and D-ribose.

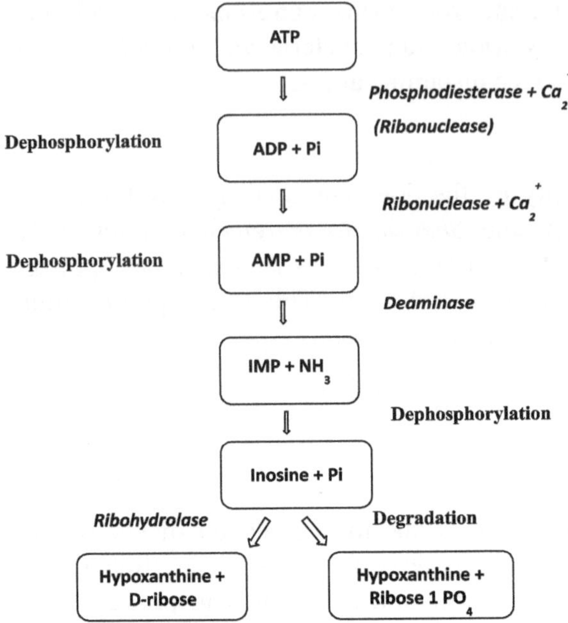

ATP degradation pathway by autolytic enzymes

3.1.3.2. Belly Bursting

Belly bursting is caused by leakage of proteolytic enzymes from pyloric caeca and intestine to the ventral part in small fatty fish like sardines and mackerels. The gut content is digested by the proteolytic enzymes into components that disrupts belly walls. The spoilage bacteria then utilize the gut components to produce gases such as CO_2 and H_2, which leads to belly bursting within short storage period.

The proteolytic enzymes present in stomach, intestine, muscle, bacteria and prey contribute for the total activity. Trypsin-like enzymes play an important role in degradation of myofibrillar proteins, but not in the solubilization of connective tissue. Collagenases are responsible for the degradation of connective tissue. The presence of animal components in the diet stimulate proteolytic activity and detritus suppresses it. So, the belly bursting is common in predatory fish than omnivorous fish. The rate of belly bursting is reduced in anchovy chilled in refrigerated seawater or iced seawater at 0°C. The dipping the fish in 15% NaCl solution for 30 min also prevents belly bursting.

3.1.3.3. Muscle Tenderization

Muscle tenderization occurs due to the action of endogenous enzymes that promote proteolysis of muscle proteins and connective tissues as well as hydrolysis of lipid. Spoilage microflora also produces enzymes that cause proteolysis, deamination, and decarboxylation, but the result in accumulation of unpleasant metabolites and loss of taste substances.

These proteases are present in muscle fiber and cytoplasm; and also, in the extracellular matrix of connective tissue surrounding muscle cell. Most of them are lysosomal and cytosolic enzymes and few are present in sarcoplasm and associated with myofibrils. Some enzymes may also originate from fish infected with myxosporidian parasites.

Two major proteinases that cause post mortem muscle tenderization and softening of fish gels are lysosomal cathepsins and calpains. Lysosomes have 13 cathepsins that play roles in protein turnover and postmortem rheological changes in muscle. Cathepsins B, D, H, L, L like and X are characterized from fish and shellfish muscles. Calpains are neutral Ca^{2+} activated proteinase involved in degradation of myofibrillar and neurofilament proteins, activation of phosphorylase b kinase and protein kinase C. Calpains are considered as the major proteolytic enzyme in postmortem muscle tenderization. The activity is controlled by Ca^{2+}, phospholipids, calpastatin, and activators. Only m-calpain is present in fish muscle. Calpains completely hydrolyze myofibrillar proteins in crustacean muscle, but only selectively cleave myofibrillar proteins into fragments in vertebrate muscle.

3.1.3.4. Black Discolouration (Melanosis)

Melanosis or "black spot" is a very common problem in crustaceans such as shrimps, crabs, and lobsters during storage in ice. It is termed as occurs as "enzymatic browning" as it is caused by an oxidative biochemical reaction catalyzed by polyphenol oxidase (PPO). In crustaceans, tyrosine is oxidized to melanin by tyrosinase (PPO) in the presence of oxygen. The melanin pigments form chiefly in cephalothorax region of crustaceans. It causes no direct harm to consumers but affects the eating quality, damages sensory features and decreases the quality, shelf life, and commercial value. Prevention is retardation of chemical reaction caused by either oxygen, enzyme, copper, or substrate by several ways.

1. Beheading, peeling and deveining, as PPO is a concentrated in blood and cephalothorax.
2. Reducing the temperature to control the enzymatic action e.g. at -18^{0}C
3. Cooking the product as increased temperature to reduce PPO activity
4. Removing the oxygen or creating inert gas atmosphere in packages
5. Using copper binding agents e.g. EDTA, as Cu is a co-factor of PPO

6. Applying irradiation e.g. low dose can reduce melanosis
7. Applying high pressure technology e.g. 900 MPa for 30 min at 45°C reduces melanosis
8. Applying dehydration process, as low water activity reduces enzymatic action
9. Decreasing the pH, as acidic medium inactivates PPO

Chemical compounds used to inhibit melanosis

1. Reducing agents such as sulfiting agents such as ascorbic acid, isomer erythorbic acid, etc. Sodium metabisulfite is most commonly used in seafood industries. *A dip treatment in 3000 ppm sodium or potassium meta bisulfite for 2 min prevents black spot formation. However, the residual level of sulfur dioxide in muscle should not exceed 100 ppm.*
2. Cysteine and Glutathione
3. Chelating agents such as phosphates, EDTA, Maltol, Kojic acid, etc.
4. Acidulants such as citric acid, phosphoric acid, etc.
5. Enzyme inhibitors such as aromatic carboxylic acids, aliphatic alcohol, anions, peptides substituted resorcinols, etc.
6. Enzymes such as oxygenase, o-methyl transferase, proteases, etc.
7. Complexing agents such as cyclodextrins

3.1.4. Physical Changes

The most common physical changes that occur in chilled fish are muscle softening, gaping and discolorations.

3.1.4.1. Muscle Softening

Exhaustive stress of capture or transportation of fish causes rapid loss of ATP and rapid onset of rigor, which eventually cause disruption and softening of the edible flesh. Softening occurs due to weakening of myocommata or connective tissue that separates myotomes. The rapid onset of isometric tension applied in stressed fish cause the muscle to pull itself apart. The rapid onset of rigor cause protein denaturation along with rapid generation of lactic acid, which contribute to the softening of muscle tissue in stressed fish.

3.1.4.2. Gaping

At the time of chilling, in fish spiked to death, rigor mortis sets in twice as fast and the ATP degradation rate is 2 to 3 times greater at 0°C than at 10°C. The rapid chilling causes **'cold shock'** in several tropical fish by accelerating the postmortem stiffening at 0°C. For example, tilapia stiffened within minutes at 0°C and but only after 7 h at 22°C. Cold shock is noticed from at 0-10°C, but not at 15°C. The stiffening of muscle (rigor) caused due to rapid chilling accelerated ATP and lactate accumulation by glycogen degradation. **Cold shock** is observed more in warm water fish. The effect of cold shock during on-board handling of fish is on the filleting yield, drip loss, and gaping of fillets. **Gaping** is a condition in which flakes separate from one another or even drop to pieces, as fish enters rigor mortis rapidly. Chilling ruptures the muscle severely by weakening the connective tissues. Low-post mortem pH due to lactic acid accumulation enhances gaping. To prevent gaping in fish fillet, it is better to keep tropical fish at ambient condition for a few hours (allowing to pass pre-rigor) before icing.

3.1.4.3. Surface Colouration

Surface coloration occurs due to enzymatic and non- enzymatic oxidation in chilled fish. Chemical groups that influence the colour of fish flesh are hemes, carotenoids and melanins. Greyness arises due to melanin. Red or red/brown ascends from blood and dark muscles.

Pink / Yellow discolouration in Squid and Cuttlefish

Pink discolouration is a common quality problem in squid and cuttlefish held in insufficient ice under stacking condition. Pink color is released from the disrupted chromatophores localized in the skin upon storage and it spreads and intensifies over the skin. Large squids are more prone to discoloration than small squids. The prevention is by deskinning the squids together with proper icing and use of mild antibacterial agent (0.1% sodium azide, NaN_3). The additional measures are immediate icing, gutting, and removal of ink sac.

Normal squid tubes

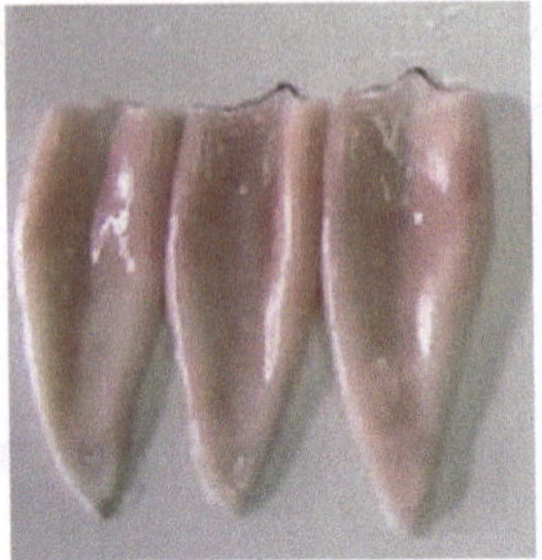

Pink discoloured squid tubes

3.1.5. Sensory Changes

Appearance of fish is the most important sensory parameter. Slime on the skin is initially watery and clear, which becomes cloudy, clotted, and discolored due to the bacterial activity. Fish mucus is secreted by globlet cells to provide the primary defense against pathogenic microbes. It contains antimicrobial peptides (AMPs), lysozymes, lectins, proteases, to provide innate immunity. It contains novel polysaccharide namely mucin, a glycoprotein. In fish held in ice, the eyes become sunken, the gills turn red brown with mucus, and the skin losses brightness and texture beyond 3 days in ice. The flesh of fresh fish is translucent, but, in stale fish it is opaque.

3.2. Quality Changes in Frozen Fish

Freezing is one of the most effective preservation methods that arrest either partially or completely the deteriorative actions of microorganisms and enzymes. Microorganisms cease to multiply below -10°C and the enzyme activity reduces at temperature below -1°C. Whole round white fish should be frozen within 2-3 days of death, flat fish within 5-6 days and small pelagic fish within 1-2 days. Shellfish, in shell on or dressed condition, should be frozen within 2-3 days. The shelf life of frozen fish depends on the initial freshness. Fish of low pH tends to deteriorate faster than those of high pH.

In processing fish for short-term storage, i.e., in ice or otherwise chilled as fillets on trays without ice, it must be decided at which stage postmortem the gutting and filleting should take place. Handling and processing of fish during rigor mortis can result in loss of quality due to gaping and decrease in fillet yield. A point of considerable interest is whether or not the fish should be gutted and filleted as soon after catch as possible. It has been shown that storage of cod as iced in gutted form rather than in the form of fillets is an effective way to prolong the freshness. Factors that affect the rate of quality decrease of the catch in any commercial operation include the delay in chilling caused by gutting and filleting as well as the hazard of bacterial contamination of the cut surfaces of the fish, related to the standards of hygiene in processing. In order to eliminate rapid autolytic changes in Antarctic krill, manifested by development of a flabby texture and excessive drip, it is necessary to separate.

Influence of Environmental Factors

The parts of the crustacean that constitute the principal source of enzymes arr to be inactivate by cooking as fast as possible after harvest. The hepatopancreas and stomach of krill are very rich in a variety of hydrolytic enzymes, including proteases, cellulases, chitinases, and lipases. The proteolytic activity of the gut content is at least five times higher than that of the whole krill. According to practical experience, the catch should not be kept onboard before processing longer than 1 h at 10°C or 3–4 h at 0–7°C. The method of choice in processing krill for high yield and meat quality is mechanical peeling on board early after catch.

3.2.1. Sensory Changes

Fresh fish is succulent with desirable odour and flavour. Frozen fish turns opaque, dull, and spongy and loss the integrity, when stored in improper conditions. White lean fish and shell fish appear opaque and yellowish, while fatty fish develops a rusty appearance. The pigments tend to fade, become dull in hue, and the gloom disappears, when dehydration occurs during frozen storage. Protein denaturation affects texture, flavor and color. Lipid–protein cross links decrease protein solubility and form colored complexes.

Frozen emperor (lean fish)

Frozen mackerel (fatty fish)

3.2.1.1. Discolourations

Colour is an indicator of the frozen fish quality, as it changes after freezing and thawing. The changes in rigor condition and muscle texture during freezing affects light reflectance properties and the visual colour of fillet. The lipid- soluble pigments get oxidized during freezing and causes “**rusting**” discolouration in flesh and develops rancid odours and flavours.

3.2.1.1.1. Yellow Discolouration

Yellow discolouration occurs below the skin of frozen cod, pomfrets, squids, and lobsters. Freezing disrupts chromatophores and releases carotenoids, which migrates to the surface fat layers and causes yellow discoloration in frozen fish.

In frozen crustaceans

Yellow discolouration in frozen shrimps and lobster meat occurs due to photo-oxidation of red astaxanthin present at the surface to a yellow colour pigment, xanthophylls, accompanied by production of off-odours and off-flavours. The prevention is through application of antioxidants, maintenance at low temperature and packaging in vacuum condition.

In squid tubes and cuttlefish fillets

Yellow discolouration in squid tubes and cuttlefish fillets occurs due to oxidation of phospholipids. The prevention is by the removal of the appendages, ink sac and gut contents, followed by washing and bleeding, immediately after catch, and storage in ice and water until freezing.

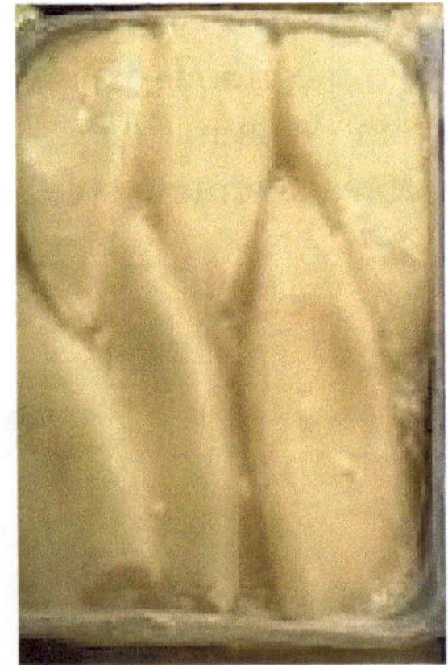

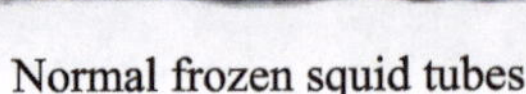

Normal frozen squid tubes Yellow coloured frozen squid tubes

In frozen pomfrets

Yellow discolouration or rusting occurs in frozen pomfrets due to oxidation of fat and formation of thick irremovable white membrane around the eyes and an undesirable odour on thawing. Bruising of fish during handling ruptures the tissue, and leads to rapid oxidation. The prevention is by glazing fish with ascorbic acid (0.1%) or a mixture of sodium chloride (0.5%) and glucose (0.5%). The thin film that forms prevent the reaction of oxygen, retard dehydration and control discolouration for 5 months.

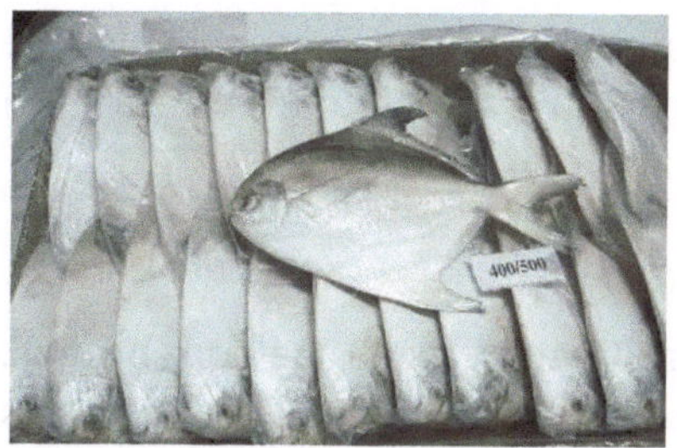

In salmon fillets

Yellow discolouration in salmon fillets occurs due to oxidation of carotenoids during freezing and thawing. A high content of conjugated double bonds exposed to oxygen causes color loss in carotenoids.

3.2.1.1.2. Green Discolouration

Green discolouration is a common defect in frozen fillets of tuna and swordfish. Fresh fish contains four forms of myoglobin namely deoxy myoglobin (purplish red), oxymyoglobin (cherry red), metmyoglobin (brown), and carboxy myoglobin (cherry red). The ordinary pink color of cooked tuna meat is due to hemochrome, derived from the reaction of myoglobin with non-heme constituents. In some occasions, the cooked tuna turns tan or tannish green colour, referred to us "green" discolouration. Cooking denatures myoglobin and oxidizes oxy, deoxy and carboxy myoglobin to yield metmyoglobin (green meat).

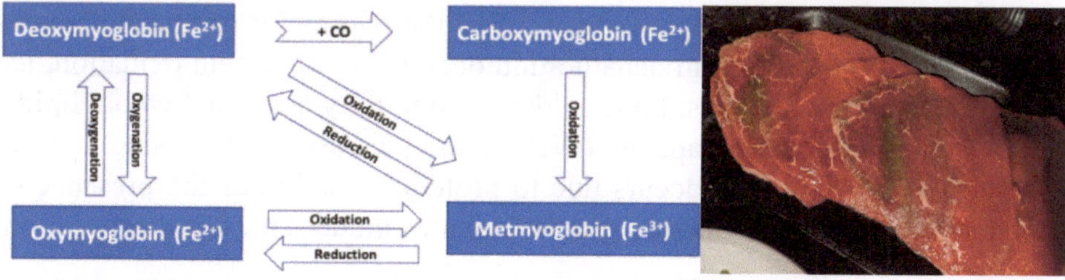

Hydrogen sulfide produced by putrefactive bacteria (*Pseudomonas mephitica, Alteromonas putrefaciens* and *Proteus vulgaris*) reacts with hemeproteins in the presence of oxygen to form sulfhemoproteins (sulfhemoglobin and sulfmyoglobin). Sour smell and off-flavour associated with greening is due to H_2S and isovaleric acid. Prevention is by proper icing, evisceration and removal of blood immediately after the catch. Discolouration can also be controlled by freezing fish at full rigor, storing at a temperature of -23°C to -27°C, and defrosting by still air at 10°C.

3.2.1.1.3. Black Spot Formation

Black spot formation or melanosis is also a problem in frozen shell on shrimps. The cause and prevention are discussed earlier under this lesson. Prevention is by proper glazing, wrapping and low temperature storage. In lobster tail, black spot is controlled by keeping them alive until the time of processing.

3.2.1.2. Flavour Changes

Fresh fish has a seaweedy odour, which is lost after freezing. The oxidation of lipids, enzymatic changes and the TMAO degradation products produce unpleasant rancid odours and flavours. The characteristic off-odours and off-flavours in frozen lean fish and shellfish are acid, bitter, turnipy, cardboardly, or musty; and in frozen fatty fish are rancid, oxidized, painty or linseed oil-like. Acid and carbonyl compounds formed due to oxidation of lipids and pigments lead to unpleasant rancid odours and flavours. Presence of blood or tissue from kidney or viscera or liver changes its flavour and texture in frozen fish mince.

3.2.1.3. Texture Changes

Frozen fish texture changes gradually from soft, moist, succulence to unacceptable firm, hard, fibrous, woody, spongy, or dry. Stale fish when frozen generates large amounts of exudate (drip). If such fish is further smoked, it gives a dull matt product instead of attractive glossy pellicle formation. The muscle tissue of frozen stale fish fails to form a gel with salt on steaming. All these textural changes occur due to protein denaturation.

3.2.2. Chemical Changes

Protein denaturation and lipid oxidation are the two major chemical changes occur in frozen fish. Protein denaturation occurs due to protein oxidation, ice crystal formation, enzymes, lipid oxidation, free fatty acids and whole lipids. It reduces water holding capacity (WHC), and increases drip causing textural changes. Lipid oxidation occurs due to protein oxidation or the presence of free fatty acids. Lipid oxidation products in turn induces protein denaturation.

3.2.2.1. Protein Oxidation

Proteins are susceptible to oxidative reactions. When proteins are exposed to oxygen, amino acids get destructed to form protein-lipid crosslinks. Protein–lipid aggregates are structured between carbonyl groups of oxidized lipids and protein molecules, to form protein-centered radicals. Modification of amino acid side chains, formation of protein carbonyls, cleavage of the peptide bonds, formation of covalent intermolecular cross-linked derivatives, loss of sulfhydryl groups are the effects of protein oxidation. The protein oxidation changes the solubility, and functionality, leads to loss of essential amino acids and decreases digestibility.

3.2.2.2. Protein Denaturation

Protein denaturation occurs in fish due to the fluctuation in the cold storage temperature, which alters myofibrillar proteins. Freezing out of water increases in salt concentration in the unfrozen phase contribute to protein denaturation. Protein aggregation occurs due to hydrophobic interactions, hydrogen bridges and the formation of covalent, non-disulfide bonds. Protein–protein bonds form high-molecular-weight polymers that are insoluble in salt solution. Secondary reaction form disulfide and formaldehyde due to low extractability and reduced functionality of proteins. Disulfide bond formation decreases the sulfhydryl group. Free fatty acids formed due to lipid hydrolysis also affect myofibrillar protein stability.

Protein denaturation leads to loss of functional properties such as WHC, viscosity, gel forming ability and lipid emulsifying capacity, etc. It causes undesirable odor, consistency, hardness and stickiness in frozen fish. The denatured protein loses its enzymatic activity. Protein denaturation is a serious problem in lean fish than lipid oxidation. Protein denaturation is lower in fish frozen in pre-rigor than in post-rigor. Protein denaturation is prevented by controlling the cold storage temperature and application of polyphosphates to protect fluid loss during thawing.

3.2.2.3. Formaldehyde Formation

In frozen fish, TMAO is broken down to dimethylamine (DMA) and formaldehyde (FA) by an autolytic enzyme, TMAO reductase on storage at as low as –8°C. Formaldehyde is an index of frozen fish deterioration. The FA deteriorates texture and functionality, mainly the extractability of myofibrillar proteins. The FA reacts with the functional groups of protein side chains and forms intra and intermolecular methylene bridges and interaction compounds with fluorescent properties. The myofibrillar proteins then denature and lead to protein aggregation. The FA accumulation is higher in dark muscles than in white muscles. The free bound water held together by FA separates out during thawing, and gives a dry and cottony texture to fish at later stages.

3.2.2.4. Lipid Oxidation

Fatty fish are more prone for lipid oxidation than lean fish. Fatty fish has a shelf life of 4–6 months at –18°C, whereas lean fish has 7-12 months. Fish lipids are oxidized to hydroperoxides in the presence of oxygen and pro-oxidants such as heme pigments and trace elements. Hydroperoxides are then decomposed into non-volatile and volatile compounds viz. hydroperoxy, hydroxy, aldehyde, epoxy and ketone compounds. Malondialdehyde and glutaraldehyde interact with amines, nucleosides and nucleic acids, proteins, amino acids and phospholipids, or other aldehydes (end-products of lipid oxidation), to form polymers causing structural and functionality changes. Hydroperoxides trigger oxidative interaction with sulfur-containing proteins leading to loss of essential amino acids and decrease the digestibility.

Oxidation of lipid and production of off-flavors in frozen fish is controlled by maintaining low cold storage temperature. Oxygen attacking unsaturated fatty acids in the fish can also be controlled by glazing or proper packaging. Glazing is used to protect against dehydration as a physical barrier between the product and oxygen in the air. It also protects against oxidation and doubles the shelf life. Packaging in oxygen impermeable flexible films extends the shelf life. Vacuum packaging or modified atmospheric packaging can be used. Addition of antioxidants also inhibits oxidative rancidity. As sodium chloride is a pro-oxidant, it should not be mixed with or applied to frozen fish.

3.2.2.5. Lipid Hydrolysis

Lipid hydrolysis occurs in frozen fatty and lean fish by the action of endogenous fish enzymes active even at –20°C. Phospholipase initiates hydrolysis of phospholipid to form free fatty acids (FFA). Increase in FFA cause rancid flavor and affect the ATPase activity. The FFA cause texture deterioration by interacting with proteins. The hydrophobic effect of FFA on proteins, and interaction of oxidized lipids with cystine-SH, $-NH_3$ group of lysine and N-terminus groups of aspartic acid, tyrosine, methionine and arginine of fish proteins, influence the solubility of myofibrillar, and sarcoplasmic proteins. The FFA oxidize more rapidly than higher-molecular-weight lipids (triglycerides and phospholipids).

3.2.3. Physical Changes

Crystallization of ice and desiccation are the major physical changes in frozen fish.

3.2.3.1. Crystallization of Ice

Crystallization of ice occurs when the temperature is lowered to about -1°C. At this time, various inorganic salts and organic components in the fluid decreases and drops the freezing temperature. The volume of the fish increases when the water converts to ice. Most part of the water freezes between -1°C to -5°C, termed as "maximum zone of crystallization". Slow freezing forms large extracellular ice-crystals, which cause textural damage due to tissue stretch, accelerate enzyme activity and increase oxidation rates during storage. Rapid freezing forms small ice- crystals, which minimize the migration of water into the extracellular space, but cause lethal intracellular ice-crystallization in living cells leading to mechanical cracking.

3.2.3.2. Freezer Burn (dehydration/desiccation)

Dehydration is a major quality loss in frozen fish. The surface and thin parts of fish turn into dry, opaque, porous and spongy structure. This condition is also termed as "Freezer burn", as it gives white patches on frozen fish meat due to severe dehydration/ desiccation during cold storage.

Wide fluctuations in cold storage temperature leads to rapid migration of water molecules and sublimation of a layer of ice crystals from the surface causing desiccation, which results in dehydration of fish. Thin parts of the fish such as fins and tail ends are more sensitive to freezer burn. It leads to weight loss and quality loss in fish, but it is safe to eat.

Dehydration can be controlled by reducing or preventing the loss of moisture (desiccation) through either coating the frozen fish in a thin layer of ice (glazing) or packaging in tightly fitting wrapper with exclusion of air. The critical limit of dehydration or water vapour transmission rate (WVTR) in frozen fish during cold storage is 50 g/m^2/day.

Glazing is the layer of ice deposited by simply dipping the frozen product in or spraying or brushing it with clean water. Glaze evaporates from the product first rather than the moisture. The glaze thickness of 0.5 to 2 mm is recommended i.e. 5% to 15% of weight. Water tends to evaporate from the product surface and condense on the interior of the package, which is termed as **"in-package desiccation".** Coatings also afford protection against dehydration.

3.2.3.3. Thaw Drip

Frozen fish that had undergone protein denaturation, when thawed loses fluid or drip known as **thaw drip**. Thaw drip consists of water-soluble proteins, vitamins, minerals and flavor-bearing components. Freezing disrupts muscle cells and releases mitochondrial and lysosomal enzymes into sarcoplasm.

Thaw drip causes a weight loss of up to 5%. Seafood that exudes a large amount of drip is usually dry and woody or tough in texture. The weight loss is mostly compensated by addition of excess initial weight. Freezing of lean fish fillets in pre-rigor compared to post-rigor increases drip loss. Thaw drip loss can be prevented by treatment with phosphates i.e. by dipping the fish fillets immediately in a 5-10% of polyphosphate solution (sodium or potassium tripolyphosphates/sodium or potassium pyrophosphates/sodium hexametaphosphate) for 1-2 min. Overtreatment results in a slimy unpleasant product.

3.2.3.4. Gaping

Gaping also occurs in frozen fish, when the connective tissue in fish muscle fails to hold the muscle blocks together and causes split or crack on the surface of a fish fillet. The gaping occurs more in fish with low muscle pH. Gaping can be controlled by chilling and freezing fish pre-rigor such that fish completes rigor stage during frozen storage and thawing. Filleting fish in pre-rigor stage allows muscle fibers to contract freely by reducing tension between muscle fibres and myocommata and thus reduces gaping.

3.2.3.5. Loosening of Head in Whole Frozen Shrimps

Loosening of head in whole shrimps is a defect at the time of catch until processing. The defect can be controlled by placing fresh shrimps in very finely crushed ice from the time of catch and processing without any delay. The use of chlorinated water at 5 ppm for washing and a dip treatment in 0.2% sodium metabisulfite for 2 min can also control the loosening of the head.

3.2.3.6. Presence of Foreign Materials and Inedible Materials

Presence of foreign materials such as flies, fiber pieces, hairs, bits of paper and excessive sand are objectionable in frozen shrimps. To control this defect, avoid peeling on floor, avoid use of pond water for washing, and stop hut-peeling to reduce sand. Proper sorting and washing controls the presence of veins, shell pieces, antennae, etc. in shrimps.

3.2.4. Microbial Changes

3.2.4.1. Bacteria

In frozen fish, intra- and extracellular ice-crystals formation cause irreversible damage to bacterial cell membranes. The increase in osmotic pressure of unfrozen solution prevents microbial growth. The frozen state of fish controls biochemical and chemical reactions. Low temperature prevents microbial growth, so microbial activity is therefore negligibly small or completely stopped in frozen fish. Most of the pathogenic bacteria are inactivated at below –10°C. Some of them grow even below –18°C.

Freezing inhibits the activity of food spoilage and food-poisoning organisms. The lethal effect of temperature on microorganisms was maximum between –4°C and –10°C rather than at low temperatures. The most sensitive bacteria are Gram-negative vegetative bacteria such as *Salmonella* and other Enterobacteriaceae. The most resistant are spores and vegetative cells of Gram-positive bacteria such as *Staphylococcus* and *Listeria*. The surviving bacteria grow and multiply in fish after thawing. *Aeromonas* spp. survive and multiply at low temperatures and produce virulent factors in frozen freshwater fish.

The maximum bacterial load in raw frozen seafood should be below 10^5/g. *Escherichia coli* in raw frozen seafood should be below 20/g and it should be absent in cooked frozen products. Coagulase-positive *Staphylococcus aureus* should be below 100/g. *Salmonella, Vibrio cholerae* and *Listeria monocytogenes* must be absent.

3.2.4.2. Parasites

Fish parasites are naturally present in sea water and fresh water. To avoid these parasites, fish needs to be cooked to an internal temperature of 70°C. Freezing of fish at appropriate temperature and duration kill the parasites. The nematode and cestode worms of public health concern are sensitive to freezing. The larvae of helminths are also inactivated by freezing. Parasites in raw or lightly preserved fish can be controlled by freezing. To kill parasitic worms, blast freezing at –35°C for 15 hours or at –20°C for 7 days is found effective. The presence of lipid in fish affect the effect of freezing to kill parasites.

3.3. Quality Changes in Cured Fishery Products

Large quantities of fish are preserved as cured products for domestic consumption in developing countries. A semi-dried product known as 'Intermediate Moisture Product - I.M.P' is now popular besides traditional salted and dried products, which has moisture up to 40-50% with good texture and flavor but shorter shelf life of 2-3 months.

The main quality defects in cured products

1. Limited shelf life due to inadequate drying
2. Heavy admixture with sand due to drying on the open beach
3. Attack by insects, fungus, mites, etc.
4. Discolouration (rusting – brown, pink and yellow)

3.3.1. Quality of Dried and Salted fish

The use of fresh fish as raw material, the use of potable water for washing and brining, the use of salt of good quality without high amounts of magnesium and calcium chlorides and the use of sodium propionate (3 ppm) in salt improves the quality of salted and dried fish. The use of raised platforms for drying, avoiding flies, proper packaging and fumigation during storage also improve the quality of products.

3.1.1.1. Moulds and Fungi

Fungi grow well in unsalted and salted dried fish having high moisture content. They include *Aspergillus niger, A. flavus, A.candidus, A.amstelodami, A.chevalieri, Rhizopus, Mucor, Penicillium spp.,* and *Polypaecilum pisce.* Moulds grow at an optimum temperature of 30-35°C with a relative humidity above 75%. Fungal growth changes colour, smell/flavour, breakdown fat and protein, and produce mycotoxins in fish. In salted fish, brownish black or yellow brown spots seen on the fleshy parts are caused by the growth of halophilic mold, *Sporendonema epizoeum.*

Prevention is by either chemical or physical methods. Chemical methods include use of preservatives, fungicides and gamma irradiation. The main preservatives are calcium propionate, potassium sorbate, sodium benzoate, parabens, sulfur dioxide, and sodium nitrite. Broad spectrum fungicides used are toxic to humans and expensive. Gamma irradiation applied have problem of recontamination, if not sealed properly. Physical methods employed are control of storage temperature, control of water activity, moisture, salt, fat, etc

3.3.1.2. Rancidity - Rusting

Rancidity occurs in salted mackerel and sardines due to the oxidation of fat imparting characteristic odour and brown colour termed as "rust". Impurities in salt and traces of copper accelerate the rancidity.

3.3.1.3. Pink /Red Discolouration

Pink/red spoilage occurs due to the growth of halophilic bacteria viz. *Halobacterium salinaria, H. cultrirubium, Sarcina morrbuae* and *S. litoralis* present in the salt on the surface as slimy pink patches. Halophilic bacteria decompose protein at 36°C and produce ammoniacal odour. Prevention is by the use of good quality salt.

3.3.1.4. Insect Infestation

Insect infestation occurs during initial drying stages and storage of dried fish. Blowflies of families, Calliphoridae and Sarcophagidae infest dried fish by laying eggs in fish, which develop into maggots that bury in the gill region. When the conditions become favorable with adequate moisture and intermittent rain, maggots develop into adult. Insect infestation causes both economic and nutritive loss to fish processors. Prevention of infestation is by maintenance of proper hygiene and sanitation, disposal of wastes and decaying matter, use of physical barriers like screens, covers for curing tanks, etc and use of heat to physically drive away the insects and kill them at 45°C

3.3.1.5. Fragmentation

Fragmentation occurs in dried fish due to denaturation and excess drying that results in breaking down of the fish during handling. Prevention is the use of fresh fish as raw material.

3.3.2. Quality Defects in Smoked Fish Products

Smoking is a method of preservation, wherein fish is subjected to brining, drying, smoking and heat treatment. The quality of smoked fish depends on the quality of raw material. If stale fish is used, the high smoking temperature volatilizes some objectionable compounds. Smoke constituents imparts characteristic yellow, golden brown or brown colors disguising minor color imperfections. Wood used for smoking upon combustion generate more than 15 polyaromatic hydrocarbons (PAHs). Among them, benzo (a) pyrene is more carcinogenic and referred to as a marker of carcinogenic PAH in smoked fish. The maximum limit set by the EU Commission Regulation is 2 ug/kg. Salt and smoke constituents also conceal strong flavors. The quality of smoked fish depends on the initial fat content. Fish with less than 1% fat content give less glossy attractive appearance, taste dry and are less succulent than those of high fat content.

The microbial quality of smoked fish depends on salt, smoke, and drying temperature. Low moisture favours the growth of moulds (*Aspergillus* spp) that originate in the wood, which turns fresh aromatic smoke flavour to weaker, blander or unpleasantly tarry. Rancidity occurs in stale smoked fatty fish. Packaging warm products leads to "sweating" and render them sticky. The use of contaminated wood or other cellulosic materials causes unpleasant resinous or acrid flavours.

Browning and **tanning** are the quality defects in smoked fish. Browning of meat occurs due to the thermal decomposition products of carbohydrates during smoke generation. Tanning effect is due to formation of secondary skin on fish during surface treatment with smoke, due the reaction of carbonyls and proteins.

Microbial spoilage in smoked fish is caused by moulds, especially xerophilic moulds, which grow at a_w of 0.65. *Aspergillus glaucus* group including *A. amstelodami, A. chevalieri, A. ruber, A. flavus, A. restrictus, A candidus* and *Pencillium* spp are responsible. Some moulds produce mycotoxins. Toxigenic strain of *A. ochraceus* produces toxins, pencillic acid and ochratoxins A. *A. flavus* and *A. parasiticus* produce aflatoxin (B1 & G1) up to 600-700 ppm in cured fish.

3.4. Quality Changes of Canned Fish Products

Canning is the heat processing method wherein fish is placed in a hermetically sealed container and processed to reduce the effect of bacterial contamination to produce a commercially sterile product. The quality of canned fish depends primarily on selection of good quality raw material without very high bacterial load. A fish in rigor is also difficult to handle in the preparatory stages of canning e.g. very fresh tuna is more liable to slow 'scorching' or darkening of the meat exposed to the headspace. Pelagic fatty fish are most preferred for canning. Fat content determines the quality of the end product and it should be between 7 and 15%. The quality changes that occur in canned are deterioration rather than spoilage.

3.4.1. Microbial Spoilage

Microbiological deteriorations occur due to insufficient heat processing during retorting or from contamination through seam. The signs of spoilage are foul smelling product, often with swelling of the can due to generation of gas (gaseous spoilage). Very rarely microbiological spoilage may not result in the generation of gas (flat sour spoilage).

1. **Flat sour (non-gaseous spoilage):** A can content is spoiled by microbial action without the formation of gas and without any external indication of spoilage known as "flat sour spoilage". It is caused by aerobic spore formers, such as *Bacillus cereus*, *B. mesentericus* and *B. vulgatus*. The product has a sour taste with or without sour odour.
2. **Gaseous Spoilage:** Cans swell or bulge due to gases produced by anaerobic or facultative anaerobic spore-formers such as *Clostridium welchii*, *Clostridium sporogenes* and gas-forming heat-resistant *Clostridium botulinum.*

Various kinds of deformity caused by spoilage are flipper, springer, soft swell and hard swell, which are evidenced by the following signs of spoilage:

1. **Flipper:** A can with normal appearance, but if one end is struck on a table, the other end becomes convex and the convexity can be pressed down again. Flipper is the initial stage of a swell, but it is also be caused by

overfilling or lack of vacuum.

2. **Springer:** A can with convex or bulging ends, which can be pressed flat again with the fingers, but will spring out again after pressure is released.
3. **Soft Swell:** A can with bulged ends, which can be moved by thumb pressure but cannot be forced back to the normal position
4. **Hard Swell:** A can with badly bulged ends, which resist pressure with the finger or even if the ends are pressed down, they spring back immediately on the release of pressure.

The chief causes of microbial spoilage in canned products are under processing, inadequate cooling, leakage through seams and pre-process spoilage.

1. **Under processing:** This causes "flat sour spoilage" or "gaseous spoilage" as a result of the activities of microorganisms surviving thermal processing.
2. **Inadequate cooling:** This spoilage occurs due to multiplication of thermophiles rapidly, if cooling cans is not done immediately after processing to a temperature of 35°C.
3. **Leaker spoilage:** This spoilage is caused by the entry of spoilage microorganisms through an opening in the improper seams formed by compressing cover hook against the folded body hook. Cooling water is the main source of microorganisms. Chlorination of process water helps to eliminate or reduce the numbers of microorganisms.
4. **Pre-process spoilage**: This spoilage occurs due to faulty canning practices; wherein bacterial development occurs in the product during the preparation. Subsequent heat processing sterilizes the pack, but the liberation of gas produced by microorganisms during the lag period before processing may cause swelling or flipping of the cans.

3.4.2. Chemical Spoilage

Chemical spoilage occurs due to the reaction of substances in the product on the metal of the can after heat processing.

3.4.2.1. Hydrogen swell

Hydrogen swell is formed due to corrosion of tin plate. Hydrogen swell cannot be easily distinguished externally from microbial swell, but can be identified only by head-space analysis and through heavily etched can interior.

3.4.2.2. Curd Formation in Canned Mackerel and Salmon

Curd formation is an appearance of clumps or layers of coagulated protein in the surface of canned mackerel and salmon. It occurs when fish are canned without pre-cooking because during heat processing, the meat coagulates, adheres to can interior and gives an unattractive appearance on opening the can. Sometimes the lacquer peels off while removing the curd. The main causes are the use of stale raw fish, badly frozen and thawed fish, inadequate brining and pre-cooking. Prevention is by soaking raw fish in 10-15% brine for 20-30 min or pre-cooking, or using tartaric acid to seal the cut surfaces or preventing the escape of fluid containing coagulable protein.

3.4.2.3. Discolourations

Many discolourations occur in canned seafood involve metal ions. For instance, blue discolouration of crab meat involves copper, whereas black discolouration in prawns relates to iron. Eels, abalone and albacore tuna suffer discolouration due to high iron content in the raw material. The discolouration progresses more when the raw material is frozen prior to canning due to the build-up of free sulfur in the tissue.

1. Black Discolouration

Black discolouration is associated with canned crabs, clams, shrimps and lobster. A black iron sulfide is formed under alkaline condition when volatile sulfur compounds that break down during processing react with iron of the tinplate. Iron sulfide is not injurious but gives an unappetizing appearance and unpleasant flavour. Prevention is by the use of parchment paper can liners or the addition of organic acids e.g. citric acid or phosphoric acid as filling medium or by dip treatment of meat. The organic acids inhibit the release of sulfur compounds and act as chelating agents. The use of can coated with special lacquer containing zinc oxide reduces black discolouration, as zinc combines with sulfide compounds to white harmless zinc sulfide. The use of parchment paper between fish and can interior and aluminum cans also prevent blackening.

2. Blue discolouration in canned crabmeat

Blue discolouration is a common defect in canned crabmeat. A blue colour copper sulfate is formed when the sulfur compounds released during processing reacts with copper present in haemocyanin, a blood pigment. The average copper level in blue meat is 2.8 mg% wet wt., which is higher than in normal meat (0.2 mg%). Prevention is partially done by cooking at a moderate temperature to coagulate the meat protein, but allowing the fluid blood pigments to drain away. Prevention can also be done by treating crabmeat before cooking in a brine containing chelating agents such as citric acid, aluminum sulfate or EDTA.

3. Green discolouration in canned tuna

Greening occurs canned tuna detracting the normal buff or pinkish tan color, due to oxidation reaction between TMAO and meat pigments such as myoglobin during heat processing in the presence of cysteine giving a bitter taste. It occurs when the TMAO level exceeds 13 mgN% or when stale fish is used. It also occurs in cooked tuna used for canning. Prevention is by addition of antioxidants or thorough bleeding after catch or avoiding the use of tuna with high concentration of TMAO and myoglobin for canning.

4. Pink discolouration in canned salmon

Red salmon (*Oncorhynchus nerka*) fetches a higher price than medium-red salmon (*Oncorhynchus kisutch*), which commands a higher price than pink salmon (*Oncorhynchus gorbuscha*). Pink discoloration occurs in canned salmon due to poor-quality raw fish or severe heat processing.

5. Brown discolouration

Brown discolouration in canned fish is due to the release of ribose from RNA in spoiling fish by riboside hydrolase. Prevention is to pre-cook the fish and decant the cook-put to remove soluble ribose or addition of *Lactobacillus pentoaceticus* that removes all ribose at 0°C in 2 days.

Brown discolouration that occurs in white pomfret and sardine is caused by the reaction of protein or amino acid with the product of lipid oxidation.

Another brown discolouration occurs in marinated fish when packed together with onion. It occurs due to a non-enzymic, carbonyl-amino browning reaction between the amino acids and 2,5-diketogluconic acid, released by bacterial action from onions.

6. Red discolouration

Red discolouration occurs in tainted canned fish due to microbial process that causes deep red color near the backbone of fish, which later fades on exposure to the air.

7. Blue discolouration in canned clam chowders

Blue discolouration in canned fish and clam chowders is due to the use of copper lined can filling machines. The chowder in contact with copper dissolves some salts, which then react with sulfides to form copper sulfide during processing and cause discolouration.

8. Black discolouration in canned clams

A dark inky black discolouration occurs in canned clams, if the dark stomach or body mass is not removed prior to canning. Prevention is to clean and wash the clams thoroughly.

9. Orange discolouration in canned tuna

Orange discolouration occurs sporadically in skipjack and albacore tuna. It is not detectable in the raw material and hence, end product inspection is the only way for its detection.

3.4.3. Physical Spoilage

3.4.3.1. Buckling of Cans

When the internal pressure during processing make ends of a can to bulge or distort, and later return back to their normal position on cooling, such condition is termed "buckling". The strain exerted seam subsequently make cans leak and allow microorganisms inside.

3.4.3.2. Leakers

Cans exude a portion of the contents, which occur due to faulty seaming, defective tin plate, internal corrosion or external rusting, buckling, gas formation by microbial action or by hydrogen gas through corrosion, external damage caused by excessively rough handling, and nail holes.

3.4.3.3. Panelling of Cans

Cans rupture or distort due to excessive external pressure causes panelling, which is an opposite of buckles.

3.4.3.4. Perforation and Corrosion

Perforation appears when the product corrodes or etches the can interior, but it is rare in canned fish. Corrosion occurs in the presence of oxygen through an electrolysis process and forms tin and iron oxides; and liberates hydrogen. Slack-filled cans are more prone to corrosion than well filled cans. Cans with low vacuum are more liable to corrosion than those with a high vacuum. Condensation of moisture or inadequate washing can also cause corrosion. Prevention is to store the product in a cool well-ventilated place with controlled storage temperature, and to cool the product thoroughly prior to stacking.

3.4.3.5. External and Internal Rusting

External rusting occurs due to sweating of cans in the storage place. Sweating occurs when the temperature of the storage place in higher than the temperature of the product or if the relative humidity of the atmosphere is high or if there is excessive variation in temperature or lack of ventilation.

Internal rusting occurs due to the combination of iron with oxygen in the presence of moisture, which is influenced by temperature and acidity or pH. Improper slack fill with excessive headspace and insufficient exhaust has oxygen. Internal rusting is not a problem in fatty fish such as salmon, tuna or sardines, as oxygen is absorbed by the product. It occurs in lean fish products such as fish roe and clams due to the presence of considerable liquid.

3.4.3.6. Honey Combing in Canned Tuna

Honey combing is a common defect in canned tuna caused due to the use of stale meat, physically damaged meat, improperly frozen and thawed meat for canning. When such meat is cooked, it gives out water and evaporates, the volume of meat contracts due to coagulation of muscle protein at the surface, the gas in the meat expands and leaves hollow spaces or perforations in the meat. On cooling, the perforations remain in the meat as small holes or air spaces, which is seen as honey comb. This condition also occurs in canned salmon and sardines.

3.4.3.7. Black Discolouration Due to Faulty Processing

Black discolouration occurs due to high processing temperature and pressure in canning. Canned clams, crabs and fish balls processed at 240°F (10 lb) give an unpleasantly dark colour than those processed at 5-8 lb pressure. Prevention is to employ pressure at less than 10 lb pressure for canning of non-acid fish products.

3.4.4. Miscellaneous defects

Some of the miscellaneous physical defects in canned seafood are underweight in canned shrimps, excessive water in oil packs and high sand content in canned mussels.

3.4.4.1. Stack-burn

Stack burn is a discolouration caused due to over processing. When cans are stacked hot into outer cartons or closely piled together, a considerable amount of heat is retained over long periods and cooking proceeds affecting both colour and flavour of the product. It is a common defect in canned clams, lobsters and salmon. Prevention is slow cooling or pressure cooling of cans before stacking.

3.4.4.2. Struvite Formation

'Struvite' is a hard, glassy crystal formed after heat processing from the natural constituents of fish. It is occasionally found in canned crustaceans, salmon, tuna and scombroids. It is often mistaken for glass by consumers. Glassy crystals are chemically calcium or magnesium ammonium phosphate, $MgNH_4PO_{4.}6H_2O$ formed during storage when pH is high. Prevention is by lowering the pH values below 6.0 with citric acid before canning or addition of chelating agent such as 0.5% sodium hexametaphosphate or sodium acid pyrophosphate.

3.4.4.3. Mush

Mush is a flabby condition occurring in some pilchards caught at the end of spawning period. A parasitic protozoan, *Chloromyxum* decomposes fish meat and turns it entirely soft during canning.

3.4.4.4. Foreign Tastes

Foreign tastes emerge when lubricating oil from filling or sealing machine drips into the cans. The reason is lack of sanitation i.e improper cleaning of machineries. Foreign tastes occur due to presence of microorganism also. Presence of odouriferous plankton viz. pteropods gives disagreeable flavors in canned salmon. Insufficiently cleaned clams give foreign tastes in product.

3.4.4.5. Undesirable Textures

The texture becomes fibrous and stringy along with salty flavour, when brined or salted fish is used for canning. The texture turns dry, chewy mouthfeel like damp cotton-wool, when frozen fish is used due to excessive protein denaturation. The texture deteriorates due to loss of more water when stale fish is used as raw material. However, the texture loss is less when fatty fish is due to the restrictive effect of lipids on water migration.

3.5. Quality Defects in Coated Fish Products

Coating refers to batter and/or breading adhering to a fish product. Coated products are tender and juicy inside and crispy outside. Coating protects juiciness of fish meat during freezing or reheating. The unit operations include portioning/forming, pre-dusting, battering, breading, flash frying, freezing, packaging, and storage.

Pre-dusting means very fine, dry, raw flour material is sprinkled on the moisty surface of fresh fish. Batter is a liquid mixture composed of water, flour, starch, and seasonings. Battering means fish portions are dipped in batter prior to cooking. Batters are of two types: adhesive and tempura. Traditional adhesive batter is a fluid consisting of flour and water. Tempura batter is puff-type special batter that forms a crisp, continuous, uniform layer over fish e.g Corn flour. Tempura batters are used with raising/leavening agents at very high viscosity levels. Tempura batter used in conjunction with coarse crumbs is a new coating process. Breading is a dry mixture of flour, starch, seasonings, coarse composition. Breading is applied to battered food products prior to frying. Pre-frying is done for a few seconds to coagulate batter prior to freezing. Important seafood coated products are breaded and battered raw and precooked fish, shrimp, fish fingers, scallops, fish balls, and fish fillets.

Coating provides moisture barrier, reduces weight loss, restricts volatile flavour loss, and restricts foreign odour pick up during frozen storage, reduces oil uptake during frying, increases the bulkiness, increases nutritional value through incorporation of nutrients in the coating, reduces costly fish content, prevents oxidation, and prolong the shelf-life of the product.

The common quality problems in coated products are:

1. Moisture loss due to dehydration during frozen storage.
2. Rapid spoilage due to microbial contamination: *Bacillus cereus, Bacillus subtilis*, and *Bacillus licheniformis* is a concern in coated products. The primary source of *Bacillus* spp. is bread crumbs. Other sources of *Bacillus* spp. are batter mix, wheat flour, and spice mix.
3. Undesirable changes due to protein denaturation and lipid oxidation
4. Undesirable colour affects the marketability: Lack of golden yellow and darker colour are not acceptable. Colour is influenced by bread crumbs, size of its granulations and cooking.
5. Loss of adhesion gives unpleasant appearance and decreases the marketability: Factors affecting adhesion are food substrate properties, cooking and batter ingredients.
6. Loss of texture affects the acceptability: Extruded crumbs gives firm to hard texture

Some other specific quality defects in coated fish products are:

1. **Voids :** Voids are bare areas on a fish portion that do not accept the batter. It happens due to high line speed, shape of the fish, absence of pre-dusting, non-adhesive surface, ice glaze or air pockets formed during application. It is rather difficult to remove it once formed due to the thick consistency of the batter, however can be controlled by removing the portion from the production line.

2. **Pillowing :** Pillowing appears as an elevated dome of batter on the product with a large air pocket beneath it. It occurs due the vapourization of water trapped under the batter during frying. As the product cools, the puffed dome collapses and creates an undesirable wrinkled appearance. Improper blending of the batter mix and high leavening levels of batter mix are the reasons.
3. **Tailings:** Batter extends beyond the product like a tail or stringer due excessive batter, which blows off during frying.
4. **Blow Off:** All the batter blows off during frying. Voids accelerate this problem. The lingering portion gives the product a dark appearance.

   Breading also blows off during frying breading. The prevention of breading blow off is by monitoring free water/ice on the surface of the substrate.
5. **Shelling:** Breading separates from substrate due to uncontrolled release of moisture.
6. **Poor Adhesion:** Breading does not adhere to the substrate. Prevention is by increasing the protein used for binding.
7. **Gummy Interface:** Gumminess in the area between surface of the substrate and breading. Prevention is the use of fully cooked product.
8. **Excessive/ insufficient pick up**: Thin or thick layer of coating. Prevention is correct management of batter viscosity.
9. **Doubles:** Fried products stick to each other. Prevention is adequate line speed during transfer and maintenance of correct batter viscosity.
10. **Belt marks:** Elongated lines on the product. Prevention is to reduce pressure of rollers in the breading operation and to have appropriate absorption of breading on the surface.
11. **Flares and tails**: Excess amount of batter causes flares and tails in the fried product. Prevention is to reduce the viscosity of batter.
12. **Ballooning:** Ballooning occurs at the time of separation of the coating material from the product after frying. It causes cracks and makes breading material fall off from the product. It occurs due to fast hardening of outer coating during pan-frying do not allow water vapor to easily escape from the product. Prevention is to adjust the viscosity and thickness of the batter and addition of fat or gums to modify porosity of the coating.

References

Badonia R., Ramachandran A. and Sankar T.V. 1988. Quality Problems in Fish Processing. Journal of the Indian Fisheries Association 18, 1988. 283-287

Horner WFA. 1997. Canning Fish and Fish Products. In: Fish Processing Technology, G. M. Hall (Ed.), pp. 119-159. Chapman & Hall

Jeyasekaran G. and Jeyashakila R. 2020. Quality Problems in Aquatic Food Products. Chapter 4. In: Quality Assurance of Aquatic Foods. pp 48- 79, Narendra Publishing House, Delhi (India)

Lu N.G. and Yerlikaya P., 2015. Quality Changes and Spoilage of Fish. Chapter 3., In: Seafood Chilling, Refrigeration and Freezing. pp. 38-57, John Wiley & Sons, Ltd., UK

Mathen, C. 1974. Quality Control in the Indian Fish Processing Industry. Fishery Technology, 11 (1): 1-16

Sungsri-in R., Benjakul S and Kijroongrojana, K. 2011. Pink Discoloration and Quality Changes of Squid (Loligo formosana) during Iced Storage, LWT - Food Science and Technology 44: 206-213.

4

Assessment of Quality Changes in Fresh, Iced and Processed Fish

Introduction

"Quality" refers to the aesthetic appearance and freshness of fish. It also involves safety aspects such as being free from harmful bacteria, parasites or chemicals. The best quality fish is the one consumed within the first few hours after postmortem. We studied in our previous chapters that in some instances, even very fresh fish in rigor mortis are unsuitable for filleting. So, slightly older fish in post-rigor are more desirable for the processors.

The assessment methods of fish quality are divided into sensory, instrumental, chemical and microbiological. Sensory method is superior and all other methods must correlate with the results of sensory evaluation. So, sensory method must be done more scientifically under controlled conditions without the effects of test environment, personal bias, etc.

4.1. Sensory Methods

Sensory evaluation is defined as the scientific discipline used to evoke, measure, analyze and interpret reactions characteristics of fish, as perceived through the senses of sight, smell, taste, touch and hearing. There are few instruments capable of measuring sensory profile viz. Instron for texture, electronic nose for odour, microscopic imaging for structure, etc.

In sensory evaluation, appearance, odour, flavour and texture of fish are evaluated by the human senses. The process involves three steps: (1) detection of a stimulus by the human sense organs, (2) evaluation and interpretation by a mental process; and (3) the response of the assessor to the stimuli. The response of stimuli varies among individuals e.g. some people cannot taste rancid flavour and some have low response to cold-storage flavour.

4.1.1. Quality Assessment of Fresh Fish

Many schemes are available for the sensory analysis of raw fish. The first modern method was developed by Torry Research Station. Later, the method was modified to give a single numerical value to a broad range of characteristics.

4.1.1.1. EU Scheme

EU scheme is the most commonly used method for quality assessment of fish in the inspection service and in the fishing, as per Council Regulation (EC) No. 2406/96. The assessment is carried out by the fish inspectors or the selling agents and a "tally" is placed in every box of fish. There are three quality levels - E (Extra), A, B, where E is the highest quality and below level B, fish is discarded. The scheme is common for all fish species (Table 4.1).

Table 4.1. EU scheme for the sensory evaluation of fatty fish freshness

Parts of fish inspected	Criteria			Not Admitted
	Freshness category			
	E	A	B	
Skin	Bright pigmentation, bright, shining Iridescent colours; clear distinction between dorsal and central surfaces	Loss of lustre and shine; duller colours; less difference between dorsal and ventral surfaces	Dull, lustreless, insipid colours; skin creased when fish curved	Very dull pigmentati on; skin coming away from flesh
Skin mucus	Aqueous, transparent	Slightly cloudy	Milky	Yellowish grey, opaque mucus
Consistency of flesh	Very firm, rigid	Fairly rigid, firm	Slightly soft	Soft (flaccid)
Gill covers	Silvery	Silvery, slightly red or brown	Brownish and extensive seepage of blood from vessels	Yellowish
Eye	Convex, bulging; blueblack bright pupil, transparent 'eyelid'	Convex and slightly sunken; dark pupil; slightly opalescent cornea	Flat; blurred pupil; blood seepage around the eye	Concave in the centre; grey pupil; milky cornea
Gills	Uniformly dark red to purple. No mucus	Less bright colour, paler at edges. Transparent mucus	Becoming thick discoloured opaque mucus	Yellowish; milky mucus
Smell of gills	Fresh seaweed; pungent; iodine	No smell or seaweed. Neutral smell	Slightly Sulphureous fatty smell, rancid bacon cuttings or rotten fruit	Rotten sour

4.1.1.2. Quality Index Method

The Quality Index Method (QIM) is originally developed by the Tasmanian Food Research unit. It is used for assessment of fresh and frozen fish. The QIM developed for redfish, sardines and flounder in the Nordic countries and Europe. A model quality assessment used to test the quality index by demerit score is given (Table 4.2).

Table 4.2. Quality assessment scheme used to identify the quality index of fish

Quality parameter	Character	Score (ice/seawater)
General appearance	Skin	0 Bright, shining 1 Bright 2 Dull
	Bloodspot on gill cover	0 None 1 Small, 10-30% 2 Big, 30-50% 3 Very big, 50-100%
	Stiffness	0 Stiff, in rigor mortis 1 Elastic 2 Firm 3 Soft
	Belly	0 Firm 1 Soft 2 Belly burst
	Smell	0 Fresh, seaweed/metallic 1 Neutral 2 Musty/sour 3 Stale meat/rancid
Eyes	Clarity	0 Clear 1 Cloudy
	Shape	0 Normal 1 Plain 2 Sunken
Gills	Colour	0 Characteristic, red 1 Faded, discoluored
	Smell	0 Fresh, seaweed/metallic 1 Neutral 2 Sweaty/slightly rancid 3 Sour stink/stale, rancid
Sum of scores		(min. 0 and max. 20)

The QIM has a score system from 0 to 4 demerit points for each sensory parameter. The description of each parameter is written as a guideline and used for evaluation. e.g. very bright skin appearance 0 demerit score, bright skin 1 and dull skin 2. The scores are given between 0-1, 0-2, 0-3 or 0-4.

Fish is inspected and the demerit point is recorded for each parameter. The scores for all the quality parameters are summed up to give an overall sensory score, called as **Quality Index**. A score of zero indicates very fresh fish, while a higher score is for spoiled fish. There is a linear correlation between the demerit score and storage life on ice. It makes it possible to predict remaining storage life of fish in ice.

4.1.1.3. Structured Scaling

For Fresh Fish

Structured scaling is a descriptive test used to determine quality and shelf-life of fish. It gives the trained panelist a scale showing several degrees of intensity for few attributes, in descriptive terms. Standards can be included at various points of the scale. The simplest method is (1) No off-odour/flavour, (2) Slight off-odour/flavour and (3) Severe off-odour/flavour, and the limit of acceptability is between 2 and 3.

For Cooked Fish

Structured scaling is also a descriptive test developed to assess cooked fish fillet of lean and fatty fish in a 10-point scale (Table 4.3). Odour, flavour and texture changes are evaluated. Statistical t-test and analysis of variance are used for comparison.

Table 4.3. Evaluation of cooked fish by structured scaling

		Grade		Score
Acceptable	No off-odour/ flavour	I	Odour/flavour characteristic	10
			of species,	9
			very fresh, seaweedy	8
			loss of odour/flavour	7
			neutral	6
	Slight off-odour/ flavour	II	Slight off-odours/flavours such as	5
			mousy, garlic, bready, sour, fruity, rancid	4
Limit of acceptability				
Reject	Severe off-odour/ flavour	III	Strong off-odours/flavours	3
			such as stale cabbage, NH_3,	2
			H_2S or sulfides	1

For Fish Products

The assessment of fish products is done by discriminative as well as descriptive tests.

1. Triangle test

The most common discriminative test used in sensory analysis of fish (ISO 4120: 1983), which indicates a detectable difference existing between two samples, A and B. The samples are presented in six different ways: ABB, BBA, AAB, BAB, ABA, BAA, to the panel members. In each coded sample, two are identical and one is different. The panel members are asked to identify the odd sample. The panel members should not be less than 12. The number of correct identifications is compared to the number expected using a statistical table, e.g., if the total number is 12, there must be 9 correct responses to get 1% significance level. It is used to identify the ingredient substitution in a product; and to select assessors to a taste panel.

2. Ranking

In a ranking test, a number of samples are presented to the taste panel. The panel members are asked to arrange them in order according to the degree to which they exhibit some specified characteristic, e.g. downward concentration of salt. Ranking can be performed more quickly with less training, and hence, used in preliminary screening.

3. Profiling

Descriptive test is very simple and used to assess a single attribute of texture, flavor and appearance. An excellent way of describing a product is done by flavor profiling. Profiling can be used for all kinds of fishery products. Quantitative Descriptive Analysis (QDA) provides a detailed description of all flavor characteristics in a qualitative and quantitative way. The panel members are given with a broad selection of reference samples and the samples are used for creating a terminology to describe the product.

A descriptive sensory analysis for fish oil using QDA is developed with a trained panel of 16 judges. Descriptive terms such as paint, nutty, grassy, metallic are used to describe the oil on an intensity scale. A moderately oxidized fish oil is given fixed scores and used as a reference, i.e. Fresh fish – 3, Amine- 1, Oily – 3, Sweet – 2, Metallic – 3, Grassy – 3, Painty – 2, Fruity – 2. The panel uses an intensity scale normally ranging from 0 to 9. Advanced multivariate analysis is used to correlate single attribute to oxidative deterioration in the fish oil. The panel consists of not less than 8-10 persons, and test is done in duplicate. The results are reported in a “spider’s web” (Fig. 4.1).

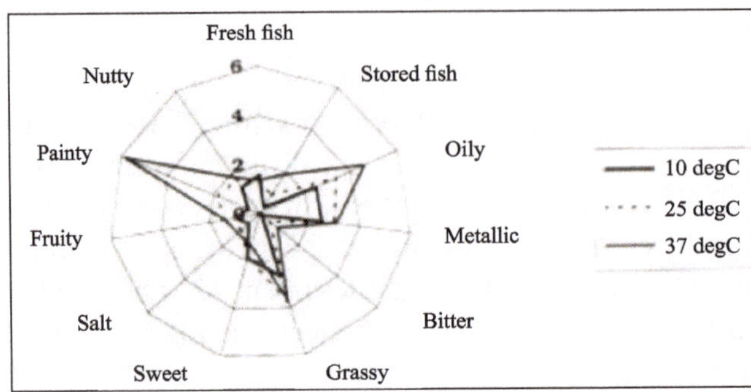

Fig. 4.1 Flavor profile of a fish oil after 2 weeks of storage at three temperatures

4.2. Physical Methods

4.2.1. Electrical Properties

Dielectric properties of skin and muscle change in a systematic way during spoilage after the death of fish. It provides means of measuring post mortem changes or degree of spoilage in fish. Several instruments are developed based on the dielectric properties of fish for the assessment of fish quality.

4.2.1.1. Torrymeter

Torry fish freshness meter or "Torrymeter" was developed at Torry Research Station in Aberdeen, Scotland. The changes in dielectric properties alter the appearance, odour, texture and flavour of fish during storage. A linear relationship exists between Torrymeter readings and sensory attributes of several fish viz. cod, Baltic herring, hake, blue whiting, flounder, mackerel, whole iced gillhead sea bream, and farmed Senegalese sole (Fig. 4.2).

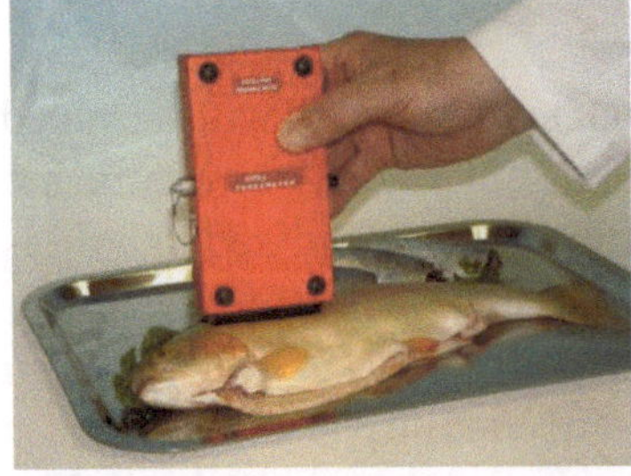

Fig 4.2 : Torrymeter

Torrymeter reading is affected if fish is washed with seawater or unwashed, as ions present in seawater interfere with dielectrical properties of skin. The presence of fat also affects the dielectrical properties. The loss of skin and muscle integrity, and deterioration of skin caused by bruising during harvest or packaging affect Torrymeter scores.

4.2.1.2. Intellectron Fischtester VI

The Intellectron Fischtester VI is similar to Torrymeter. It measures the electric properties like resistance, conductivity and capacitance of the fish flesh. The electric properties of fish change after death due to the disruption of cell membranes by autolysis. It is suitable for whole fish and fillets with skin. Mechanical abuse and freezing affect the readings. It gives information on the period of fish held in ice and remaining period in ice. It is an objective test to assess the state of freshness/spoilage across the fish chain.

4.2.1.3. RT-Freshtester

The RT-Freshtester also works on dielectrical properties of fish. The reading decreases with storage time. It allows automatic grading of 60-70 fish per min over 4 channels. It is unsuitable for frozen or thawed fish and partly frozen viz. super chilled fish and fish chilled in RSW.

4.2.2. pH

Postmortem pH of fish varies from 5.5 to 7.1, depending on season, species and other factors. A low pH is an indication of stress at the time of harvest, caused by the depletion of energy reserves, mainly glycogen with the production of lactate. A low pH causes a decrease in water binding capacity of myofibrils. A low pH promotes oxidation of myoglobin and lipids. The pH is measured with a pH-meter by placing the electrodes either directly into the flesh or into a suspension of fish flesh in distilled water.

4.2.3. Texture Analysis

Texture is an important property of fish muscle, whether raw or cooked. Fish muscle becomes soft or mushy as a result of autolytic degradation or tough as a result of frozen storage. Autolytic degradation takes place when the indigenous proteases in fish begins to break down the proteins after harvesting, during processing, improper handling, storage and cooking. Numerous mechanical methods are available to measure texture of fish. The common texture parameters are hardness, springiness and chewiness. Hardness in the most important parameter.

4.2.4. Electronic Nose

Odour is the main indicator of fish freshness. Electronic nose, also called as Freshsense was developed and distributed by Element-Bodvaki in Ireland. It is a rapid and non-destructive method used to measure volatile compounds, including spoilage odours in seafood. It is a closed, static sampling system with electrochemical gas sensors, which are sensitive to volatile compounds (Fig.4.3).

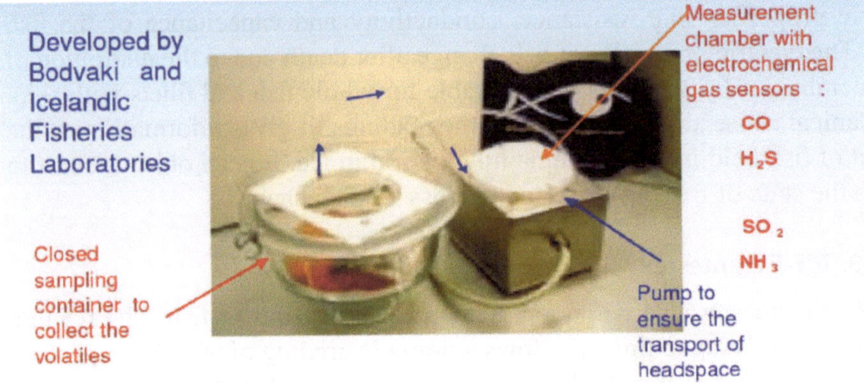

Fig. 4.3 : Freshsense, an electronic nose

The most important chemical compounds responsible for fresh fish odours are long chain alcohols and carbonyls, bromophenols, and N-cyclic compounds. During storage of fish, short chain alcohols and carbonyls, amines, sulfur compounds and aromatic, N-cyclic and acid compounds are produced by microbial activity and lipid oxidation. The concentration of these chemical compounds is related to the degree of spoilage.

Different electronic noses are available with metal-oxide semi-conductor gas sensors, electrochemical sensors (CO, H_2O, NO, SO_2 and NH_3), thickness shear mode quartz resonators, semiconductor dimethylamine (DMA) gas sensor, and prototype solid state-based gas sensor. All these sensors are known as "Fish noses". There is a good correlation between the response of CO sensor and QIM scores for both air and MAP stored fish.

Fish freshness is also evaluated by computer screen photo-assisted technique (CSPT) based gas sensor array. It is based on a computer screen, easily programmed to show millions of colors combining wavelengths in the optical range. Optics based electronic nose relies either on absorbance or fluorescence, while CSPT evaluates both effects. Data analysis determines the relation between sensor output patterns and properties. The most frequently used methods are artificial neural works (ANNs), chemometric analysis such as principal component analysis (PCA) and partial least square regression (PLS-R).

4.2.5. Near Infrared Reflectance Spectroscopy

Near infrared reflectance (NIR) spectroscopy has the ability to measure numerous samples within a short time on/at line. It is non- destructive, easy to handle and requires little training of operators. It is applied for determination of fat, water and protein contents in fish; free fatty acids in fish oils; water holding capacity of thawed fish; differentiation of fresh and frozen thawed fish; and quality assessment of fish caught by longline and gillnet.

4.3. Biochemical Methods

The biochemical methods for the assessment of fish quality are is to set quantitative standards i.e. the establishment of tolerance levels. They are more reliable and accurate, since they eliminate personal bias. Sensory methods identify products of very good or poor quality, while biochemical methods resolve issues regarding products of marginal quality. In addition, biochemical methods replace more time-consuming microbiological methods. The chemical compound to be measured should either increase or decrease due to microbial spoilage or autolysis. The compounds must not get affected by processing. Some of the most useful biochemical indicators of fish quality are given below:

4.3.1. Amines - Total Volatile Basic Amines

Total volatile basic amine (TVB) is one of the most widely used measurements of fish quality. It includes the measurement of TMA (produced by spoilage bacteria), DMA (produced by autolytic enzymes during frozen storage), ammonia (produced by the deamination of amino-acids and nucleotide catabolites) and other volatile basic nitrogenous compounds associated with fish spoilage. The European Commission (Council Regulation No. 95/149/EEC of March 1995) on fish hygiene specifies that if the organoleptic examination indicates any doubt on the freshness of the fish, TVB-N should be used as a chemical check. Therefore, it is not good indicator of fish freshness, but is a good spoilage index.

The level of TVB-N in freshly caught fish is generally between 5 and 20 mgN/100g muscle. The levels of 30-35 mg N/100 g muscle are considered as the limit of acceptability for ice stored cold water fish. The TVB-N is also useful to measure the quality in cephalopods (squid), industrial fish for meal and silage, and crustaceans.

The TVB-N analysis depends upon either steam distillation of volatile amines or Conway microdiffusion of an extract. The EC reference method for TVB-N determination, involving preliminary deproteinization with perchloric acid followed by steam distillation of fish with the addition of magnesium oxide. The direct steam distillation method is recommended as a rapid routine method.

4.3.2. Ammonia

Ammonia is formed by the bacterial deamination of proteins, peptides and amino- acids. It is also produced in the autolytic breakdown of adenosine monophosphate (AMP) in chilled fish products. Two methods are available to identify ammonia. The first one involves the use of the enzyme, glutamate dehydrogenase, NADH and alpha-ketoglutarate. The second involves only the use of glutamate dehydrogenase in test. The test strip changes colour when placed in contact with aqueous extracts containing ammonium ion. Ammonia is an excellent indicator of squid quality. It predicts the latter stages of quality in finfish. It is an objective quality indicator for fish.

4.3.3. Trimethylamine (TMA)

Trimethylamine is a pungent volatile amine associated with the typical "fishy" odour in spoiled fish. The TMA is formed in spoiling fish by bacterial reduction of trimethylamine oxide (TMAO) by TMAO reductase. The TMAO is naturally present in fish to aid in osmoregulation. Its content varies with species, age, fish size, time of year and environmental factors. Seawater fish contain 1-100 mg TMAO/100 g, whereas freshwater fish contain only 5-20 mg/100g.

The TMA content does not correlate with bacterial numbers, due to the presence of small numbers of "specific spoilage" bacteria (SSO), which are relatively present in small proportion but capable of producing large amounts of TMA. One SSO, *Photobacterium phosphoreum,* generates approximately 10-100 folds the amount of TMA than that produced by the more common SSO, *Shewanella putrefaciens.* The TMA is not a good indicator of pelagic fish quality but is useful as a measure of the eating quality of marine demersal fish. The correlations between TMA level or TMA index and eating quality is excellent.

The TMA is used as a spoilage indicator and not as an index for freshness. The TMA production is not significant during the early stages of chilled storage of fish, but it appears after 3 or 4 days, parallel to the bacterial proliferation. Fresh fish has a very low amount of TMA (<1.5 mg TMA/100 g) but the values increase during spoilage. The fish is considered unacceptable when TMA level is higher than 30 mg/100 g. Many analytical methods are available for the measurements of TMA, DMA or TVB-N contents such as steam distillation, Conway microdiffusion and titration, colorimetric, photometry, HPLC, Gas chromatography, injection gas diffusion, biosensor using flavin containing monooxygenase type-3 and solid-state sensors based on bromocresol green.

4.3.4. Dimethylamine (DMA)

Fish contain an enzyme, TMAO dimethylase (TMAOase), which converts TMAO into DMA and formaldehyde (FA) during chilled or frozen storage, when bacterial growth is inhibited. These products cause severe quality changes in fish during prolonged frozen storage. The FA induces toughening of fish proteins leading to protein denaturation. The DMA is most commonly used to assess the quality of frozen-stored fish rather than FA, because much of the FA is bound to the tissue and not extractable.

The DMA is associated with membranes in the fish muscle and hence, rough handling and temperature fluctuations enhance its production. The amount of DMA produced depends on species, the storage temperature, and time. The DMA has little or no effect on the flavor or texture of the fish. It is an indirect indicator of protein denaturation. It is hence used as a spoilage index in frozen fish. The most common method for analysis is a colorimetric determination in deproteinized fish extracts, but chromatographic methods including GC, and HPLC are more specific.

4.3.5. Formaldehyde

The presence of FA in fish is generally considered as nontoxic, but it can react with a number of chemical compounds such as amino acid residues, terminal amino groups and low molecular weight compounds and lead to denaturation and cross linking of proteins. It reduces the solubility of myofibrillar proteins. It is used as a spoilage index, especially in frozen gadoid fish. Now-a-days, FA is indiscriminately added to fish as an additive by fresh fish traders to extend the shelf life in ice. The added FA can cause severe health issues like irritation in eyes, skin, bronchitis, and cancer on prolonged exposure. Indian government has set a limit of 4 ppm for freshwater fish and 100 ppm for marine fish. Techniques used to assess formaldehyde are spectrophotometric, chromatographic and test kits.

4.3.6. Biogenic Amines

Biogenic amines are generated by microbial decarboxylation of specific free amino acids in fish. The most significant biogenic amines produced in fish are histamine, putrescine, cadaverine, tyramine, tryptamine, 2-phenylethylamine, spermine, spermidine and agmatine. Histamine is derived from histidine, cadaverine from lysine, tyramine from tyrosine, putrescine from arginine, tryptamine from tryptophan, and 2-phenylethylamine from phenylalanine. Putrescine is also an intermediate product in a metabolic pathway that produces spermidine and spermine. Biogenic amine formation depends on species, free amino acid content, presence of decarboxylase positive microorganisms, moment of capture and stomach contents in fish at the time of death.

Among the biogenic amines, histamine is associated with incidences of scombroid poisoning caused due to ingestion of tuna, mackerel, and mahi-mahi (dolphinfish). Putrescine and cadaverine are reported to enhance the toxicity of histamine. Consumption of fish containing these toxic amines has toxicological effects, which are more severe in sensitive consumers, who have less mono- and diamine oxidases (MAO & DAO), the enzymes responsible for its detoxification. The legal limit set for histamine is 5 mg/100g by USFDA and 20 mg/100g fish by the EU.

Techniques used to analyze biogenic amines include thin layer chromatography (TLC), spectrofluorometric, HPLC, GC, capillary zone electrophoresis, enzymatic and use of biosensors. The HPLC method is most performed because of its sensitivity, reliability and reproducibility.

4.3.7. Nucleotide Catabolites

Rigormortis causes death stiffening of muscle in postmortem fish. This process breakdown adenosine triphosphate (ATP), which is the main energy source for metabolic activity. There is good correlation between nucleotide degradation products and loss of freshness. The nucleotide breakdown occurs as a result of autolytic enzymes and bacterial action. The initial stage of the process is catalyzed by endogenous enzymes quickly, leading to the formation of adenosine diphosphate (ADP), adenosine monophosphate (AMP), and inosine monophosphate (IMP). The oxidation of hypoxanthine (Hx) to xanthine and uric acid is slower, as the result of either endogenous enzyme or microbial activity. The IMP is associated with fresh fish flavour, whereas inosine and Hx reflect poor quality.

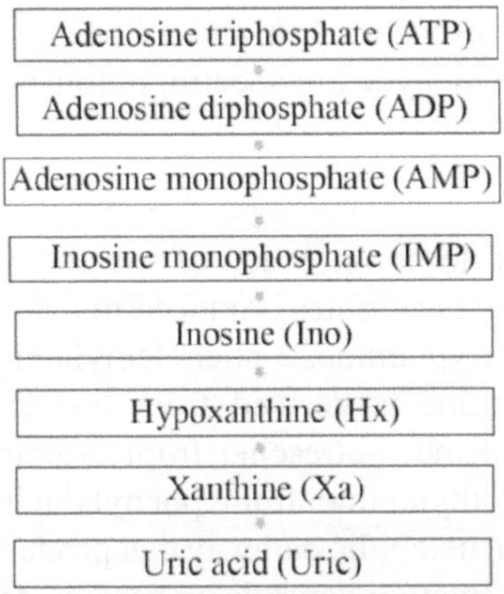

Adenosine triphosphate (ATP)

Adenosine diphosphate (ADP)

Adenosine monophosphate (AMP)

Inosine monophosphate (IMP)

Inosine (Ino)

Hypoxanthine (Hx)

Xanthine (Xa)

Uric acid (Uric)

The concentration of ATP and its breakdown products is used as indicators of freshness in fish. The concentration of each intermediate catabolite rises and falls as spoilage progresses. So, it is not possible to measure a single catabolite and hence, it is recommended to analyze the complete nucleotide catabolite profile. Saito (1959) proposed a K value as the biochemical index for fish quality assessment. The K value includes intermediate breakdown products, since adenosine nucleotides are almost converted to IMP within 24 h postmortem.

$$K\,(\%) = \frac{\text{Ino} + \text{Hx}}{\text{ATP} + \text{ADP} + \text{AMP} + \text{IMP} + \text{Ino} + \text{Hx}} \text{x}100$$

4.3.8. Ethanol

Ethanol is used as an objective indicator for seafood in early days. Ethanol is derived from carbohydrates *via* anaerobic fermentation (glycolysis) and/ or deamination and decarboxylation of amino-acids such as alanine, which is a common metabolite of a variety of bacteria. Ethanol is heat-stable and hence used to assess the quality of canned fish products. The commercial enzyme test kits are the simplest and most reliable methods for measuring ethanol.

4.3.9. Oxidative Rancidity

Fresh fish is prone for rapid deterioration with limited shelf life, whereas frozen fish state lasts for several months without severe quality changes. The only limiting factor in frozen lean fish is denaturation of proteins, which gives a dry and firm texture. In frozen fatty fish, enzymatic and non-enzymatic lipid oxidation occurs during processing and storage and leads to oxidative rancidity. Fish has highly unsaturated lipids, which under chilled/ frozen conditions undergo oxidation. The oxidation compounds interact with proteins and lead to protein denaturation, nutritional losses and modification of proteins. The development of off taste and off odour is defined as rancidity. The factors affecting the development of rancidity are degree of unsaturation, the presence of antioxidants, pro-oxidants, moisture content, oxygen availability, temperature, and degree of exposure to light.

Autooxidation of unsaturated fatty acids proceeds in three steps: initiation, propagation and termination. Light and heat are the initiators, which convert fatty acids (RH) to free radicals (initiation) and free radicals react with oxygen to produce peroxide radicals (ROO*). The peroxide radical attack another lipid molecule RH, and lead to the formation of hydroperoxide (ROOH) and new free radical R* (propagation). Peroxides are not stable and hence, they break down to aldehyde, ketones and alcohols, which are the volatile products that cause off flavor and off taste in fish. The number of reactive compounds increases gradually, and the quantity of radicals and peroxides decreases forming stable deterioration products (termination). Oxidation products are unstable and they react with proteins, peptides, free amino acids and phospholipids to produce interaction products. The presence of excess free radicals and peroxides cause destruction of essential fatty acids and vitamins A, C, E and B_6, B_2 and B_5. Peroxides also react with proteins and decrease the nutritional value. Peroxides also destroy pigments, produce toxins, and cause off flavour/odours.

The major chemical indicators of oxidative rancidity are anisidine value (AV), peroxide value (PV), TOTOX value and thiobarbituric acid-reactive substances (TBARS). The PV measures the primary lipid oxidation products (hydroperoxides) and is used as an index of the earlier stages of oxidation. The AV and TBARS values measure the secondary products of lipid oxidation (aldehydes and ketones). Analysis of oxidation interaction products by fluorescence detection is a quality index to assess quality differences during frozen storage of fish. NIR spectroscopy is used to monitor both oxidation and hydrolytic degradation of lipids in fish oils.

4.3.9.1. Peroxide Value

Peroxide value measures primary products of lipid oxidation, which break down to secondary products of oxidation or react with proteins. The hydroperoxides formed are odor- and flavor-less compounds. The PV is analyzed either by iodometry or spectrophotometry with thiocyanate. The methods use the oxidation potential of peroxide to oxidize iodide to iodine or to oxidize iron (II) to iron (III) thiocyanate to determine the concentration. The limit for PV by iodometry should not exceed 10-20 meq/kg fish fat. The spectrophotometry method gives 1-1.5 times higher PV than iodometry.

4.3.9.2. Thiobarbituric Acid Value Reactive Substances (TBA-RS)

The secondary oxidation products comprise aldehydes, ketones, and short chain fatty acids. They give very unpleasant odors and flavors, and in combination produce fishy and rancid character associated with oxidized fish lipid. A measure of TBA-RS is obtained when the aldehydic products react with TBA to form a reddish coloured product, which is determined by spectrophotometry and reported as micromoles malonaldehyde present in 1 g of fat. Headspace analysis of the volatile oxidation products can be done by GC. The guideline values given for TBA-RS is 1-2 μmol MDA-equivalent per g fat or 10 μmol MDA-equivalent per 1 kg fish.

4.3.10. Hydrolytic Rancidity

Lipid hydrolysis occurs due to enzymic hydrolysis of fat into free fatty acids (FFA) and other products. The FFA gradually increase during frozen storage of fatty fish and lead to hydrolytic rancidity. The FFA interact with myofibrillar proteins and promote protein aggregation affecting the texture and functional property (solubility).

4.4. Microbiological Methods

Microbiological evaluation is done to evaluate the presence of bacteria or organisms of public health significance in fish. It gives information on eating quality and freshness of fish. Conventional microbiological methods are laborious, time-consuming, costly and require skill in execution and interpretation of the results. Various rapid methods are now developed and automated procedures are used when the samples are in large numbers.

4.4.1. Total Aerobic Counts

Total aerobic count (TAC) or Standard plate count (SPC) represents the total number of bacteria capable of forming visible colonies on a culture media at a given temperature. It is a good indicator of the sensory quality or expected shelf life of the product. When the total count exceeds $10^5 - 10^6$ cfu/g, the fish is rejected. There is no correlation between the total counts and presence of any bacteria of public health significance.

Plate count agars (PCA) are widely used for determination of total counts. The incubation is generally at 35-37°C for 2 days. To examine psychrotrophs, pour plating followed by an incubation at 25°C for 3-4 days is more appropriate. To examine products having the psychrophilic *Photobacterium phosphoreum*, surface plating followed by incubation at maximum 15°C. To ease the determination of total counts, ready-made Redigel and Petrifilm™ SM are now available. They are dried agars with a gelling agent, to which the sample is added directly. All agar-based methods have a drawback of lengthy incubation period.

Direct microscopic examination is a rapid method for estimating bacterial count in fish. Phase contrast microscopy is used such that one cell per field of vision equals approximately 5-10^5 cfu/ml at 1000X magnification. Direct epifluorescence filter technique (DEFT) has got wide acceptance, in which cells are stained with acridine orange and detected by fluorescence microscopy. Microscopic methods are very rapid but lacks sensitivity.

Bacterial numbers are also estimated by measuring the amount of bacterial adenosine triphosphate (ATP) or by measuring the amount of endotoxin produced by Gram-negative bacteria by the *Limulus* amoebocytes lysate (LAL) test. The former is very rapid but difficulties exist in the separation of bacterial and somatic ATP.

Several methods such as microcalorimetry, impedance, conductance and capacitance, and dye reduction are used for rapid estimation of bacterial numbers. In microcalorimetry, the heat generated by the sample is compared with a sterile control, whereas in impedance, conductance and capacitance measurements, the change in electrical properties is compared to a sterile control. The change in electrical properties occurs due to the growth of microorganisms. It is used for the rapid estimation of total bacterial counts, coliforms, and *Salmonella* spp. In impedance measurements, at a time point (detection time - DT), bacteria grow to a population of approximately 10^7 CFU/mL or higher and cause an accelerating change in impedance (or conductance) in the growth media. The decrease in impedance (or increase in conductance) is due to the breakdown of the substrate molecules to smaller molecules (e.g., acids), which have more charges than the substrate itself. A comparison of the different methods used for the determination of total counts in seafood is given (Table 1.4).

Table 1.4 : Methods for determination of the total bacterial counts in seafood

Method	Temperature,°C	Incubation	Sensitivity, cfu/g
Plate count or Iron agar	15-25	3-5 days	10
"Redigel"/"Petrifilm™ SM"	15-25	3-5 days	10
Microcolony-DEFT	15-30	3-4 hours	10^4-10^5
DEFT	-	30 min	10^4-10^5
ATP	-	1 hour	10^4-10^5
Limulus lysate test	-	2-3 hours	10^3-10^4
Microcalorimetry/ Dye reduction/ Conductance/Capacitance	15-25	4-40 hours	10

Modern microbiological techniques such as polymerase chain reaction (PCR), reverse transcriptase PCR (RT-PCR), antibody techniques such as enzyme-linked immunosorbent assay (ELISA), enzyme-linked immunomagnetic chemi-luminescence (ELIMCL), and oligonucleotide probes are available, which can give results in 1 day or even less. All these methods have limitations in quantitative analyses, as they lack in sensitivity, and are costly.

4.4.2. Spoilage Bacteria

The total number of bacteria on fish rarely indicates sensory quality or expected storage life in ice. Bacteria that cause spoilage in fish are specific spoilage organisms (SSOs), and they are mostly hydrogen sulfide producing bacteria. A nutrient rich Iron Agar is used for determination of SSOs. Peptone-rich substrates containing ferric citrate are used for detection of H_2S-producing bacteria such as *Shewanella putrefaciens,* which forms black colonies due to precipitation of FeS. Members of family Vibrionaceae cause spoilage at ambient temperature also form black colonies on an iron agar. No selective medium exists for *Pseudomonas spp.* that spoil some tropical and freshwater fish or for *Photobacterium phosphoreum* that spoil packed fresh fish. A conductance-based method for specific detection of *P. phosphoreum* is available. Specific antibodies or gene probes are available to study the presence or absence of spoilage bacteria, S. *putrefaciens*.

The numbers of SSOs and the concentration of their metabolites are objective quality indicators for determination of shelf life of fish It is possible to predict shelf life of a fish based on initial numbers and growth of SSO. Mathematical models are established for the growth of spoilage bacteria such as *Photobacterium phosphoreum, Shewanella putrefaciens, Brochothrix thermosphacta, Listeria monocyotgenes, and Clostridium perfringens,* which correlate with remaining shelf life of product.

References

Cheng J.H., Sun D., Zeng X. and Liu D. 2015. Recent advances in methods and techniques for freshness quality determination and evaluation of fish and fish fillets: a review. Crit. Rev. Food Sci. Nutr. 55 (7), 1012–1225

Howgate, P. 2010. A critical review of total volatile bases and trimethylamine as indices of freshness of fish. Part 1. Determination. Elec. J. Env. Agri. Food Chem. 9, 29–57

Huss H.H., Ababouch L. and Gram L. 2003. Assessment and Management of Seafood Safety and Quality. Food and Agriculture Organization of the United Nations, FAO Fisheries Technical Paper 444. P 239

Ndraha N. 2017. Fish Quality Evaluation Using Quality Index Method (QIM), Correlating with Physical, Chemical and Bacteriological Changes During the Ice-Storage Period: A Review. In: Isnansetyo A., Nuringtyas T. (eds) Proceeding of the 1st International Conference on Tropical Agriculture. Springer, Cham. https://doi.org/10.1007/978-3-319-60363-6_18

5

Pre-requisites to HACCP

Introduction

Quality is very important in any production and operation system. It ensures whether the products produced or the services provided are of the right quality in right quantity at right time. Quality management must ensure proper quality of the product to survive in the market and expand its market. Quality of the product must be given importance to avoid quality failures that cause serious human inconveniences and wastage of money.

5.1. Definitions and Terminologies

5.1.1. Quality

We say a product is of satisfactory quality, if it satisfies the consumers. According to C. D. Lewis, a British writer, quality is "*an asset which is offered to the potential customer of that product or service*". The International Organization for Standardization (ISO) has put up a more comprehensive definition of quality as *"the degree to which a set of inherent characteristics fulfils requirements"* (ISO 9000: 2005 – 3.1.1). In general terms, 'quality' is defined as *the composite of characteristics that differentiate individual units of a product to determine its acceptability by the customer*. Quality characteristics are nothing but elements that makes a product fit for use. Quality characteristics are further classified into categories called as parameters.

5.1.2. Total Quality Management

Total Quality Management (TQM) is an organization's management approach, centered on quality. It improves the effectiveness and flexibility of a company or an organization. TQM is defined as "*the system of establishing defect prevention actions and attitudes with a company or an organization on a permanent basis for the purpose of assuring conforming products or services directed at customer satisfaction*". TQM is based on the participation of all its members in decision making, which is aimed for long-term success, through customer satisfaction and benefits to the members of the company or organization and to the society.

5.1.3. Quality Assurance

The term, Quality Assurance (Q.A) gives an emphasis on quality at the design stage of the products, processes, and procedures, including in the selection of manpower and their training. According to ISO 8402, Quality Assurance means *"all those planned and systematic actions necessary to provide adequate confidence that a product or service will satisfy given requirements for quality"*. It is a strategic management function to establish policies, to adapt programs to meet established goals and to provide confidence that these measures are applied effectively. Quality management (QM) is often used interchangeably with Quality Assurance.

5.1.4. Quality Control

The term, Quality Control (QC) is defined as *"the operational techniques and activities that are used to fulfill requirements for quality"* Any Quality Control program is based on the periodic inspection of products at various stages of production, followed by feedback on results and further corrections. So, Quality Control is a tactical function which carries out the program established by Quality Assurance.

5.1.5. Total Quality Control

Total Quality Control (TQC) refers to a total commitment to quality in all functional areas of work. It utilizes behavioral techniques such as quality circles (QC) and zero-defect program. TQC is not restricted to product quality but also on the quality of all business processes.

5.1.6. Good Manufacturing Practices (GMP)

Good manufacturing practices are the procedures for a particular manufacturing operation, which is considered the best available by the practitioners and the experts.

5.1.7. Good Hygienic Practices (GHP)

Codex defines "food hygiene" as "all conditions and measures necessary to ensure the safety and suitability of food at all stages of the food chain" (CAC, 2001). The term "Good Hygienic Practices (GHP)" refers to all practices regarding the conditions and measures necessary to ensure the safety and suitability of food at all stages of the food chain. The terms, GMP and GHP, basically cover the same procedures and are often used interchangeably.

5.1.8. Sanitation Standard Operating Procedures (SSOPs)

Sanitation standard operating procedures (SSOPs) are documented GMP available for hygiene and sanitation, and are required to meet the regulatory requirements for food control in the USA. The SSOP are equivalent to GHP.

5.1.9. Pre-requisite Programs

Pre-requisite program is equivalent to GHP, and also, GMP and SSOP. Various definitions for the pre-requisite programs are proposed by international organizations. The World Health Organization defines pre-requisite program as "*practices and conditions needed prior to and during the implementation of HACCP, which are essential for food safety*" (WHO, 1999). The Codex says *prior to the application of HACCP to any sector of the food chain that sector should be operating according to the Codex General Principles of Food Hygiene, the appropriate Codex Codes of Practices, and appropriate Food Safety Legislation* (CAC, 2001). The National Advisory Committee on Microbiological Criteria for Foods defines pre-requisite program as "*procedures, including GMP that address operational conditions providing the foundation for the HACCP system*" (NACMCF, 1998)

5.2. Pre-requisite Programs for HACCP

Pre-requisite programs are steps or procedures that control the operational conditions within a food establishment and promote environmental conditions that are favorable for the production of safe food, which include Good Manufacturing Practices (GMPs) or Good Hygienic Practices (GHPs) and Specific Standard Operating Procedures (SSOPs). Pre-requisite programs are essential for any food operation prior to the implementation of the HACCP system. Pre-requisite program is a facility-wide program rather than process or product specific program. Pre-requisite program helps to reduce hazards. Fig 1 shows the integrated approach in food safety and quality.

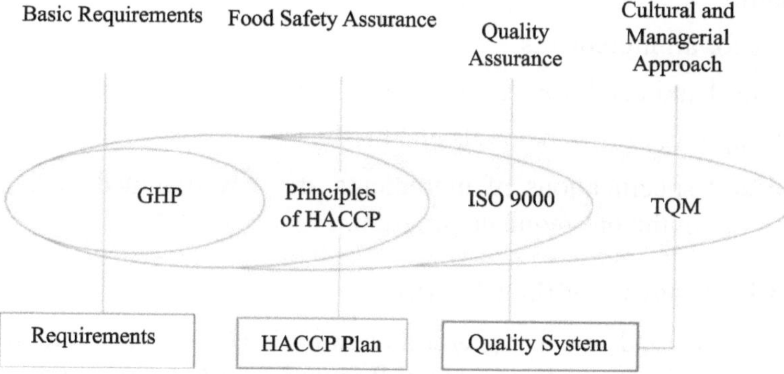

Fig 5.1 : Food Safety and Quality Management System

5.2.1. Pre-requisite Program of the Codex Alimentarius Commission

The basis for developing and implementing GHPs are the "Recommended International Code of Practice – General Principles of Food Hygiene (CAC/

RCP-1,1969, Revision 2003) – HACCP systems and Guidelines for its Application" and International Code of Practice for Fish and Fishery Products (CAC/RCP 52-2003, Revision, 2008). The following aspects are included in the pre-requisite program.

- Fishing and harvesting vessels design and construction
- Facility design and construction
- Design and construction of equipment and utensils
- Hygiene control program
- Personal hygiene and health
- Transportation
- Traceability and recall procedures
- Training

These two Codex codes of practices are used by many countries, including India for their fish and seafood hygiene regulations, by most seafood trade associations and companies.

5.2.2. Pre-requisite Program of the NACMCF

The National Advisory Committee on Microbiological Criteria for Foods (NACMCF) is an advisory committee under the U.S. Department of Agriculture (USDA). In the NACMCF common pre-requisite program, in addition to the above points listed by Codex, it also contains,

- Supplier control
- Specifications for all ingredients
- Chemical control and conditions for receipt
- Storage and shipping of raw materials and products.

Supplier control and specifications of ingredients are now included in the HACCP plan and not in the pre-requisite program.

5.2.3. Pre-requisite Program of the US-FDA

As per United States Food and Drug Administration's (US-FDA) seafood HACCP regulation (21 CFR- Code of Federal Regulations 120.6), processors are required to have key sanitary conditions written as Sanitation Standard Operating Procedures (SSOPs) (FDA, 1995), to correct unsanitary conditions and practices and to maintain sanitation control records. Sanitation control procedures are an integral part of the seafood HACCP regulation, but not of the HACCP program.

The SSOP must address the following eight conditions and practices:

- Safety of water and ice Condition and cleanliness of food contact surfaces
- Prevention of cross contamination from unsanitary objects to food
- Maintenance of facilities for personal hygiene
- Protection of food and food contact surfaces from adulteration
- Proper labelling, storage and use of toxic compounds
- Control of employee health conditions
- Exclusion of pests

The written SSOP plan should include sanitation concerns, controls, in-plant procedures and monitoring requirements. The title 21 CFR Part 117 outlines **current Good Manufacturing Practices (cGMP),** Hazard Analysis and Risk based Preventive Controls for Human Food. Title 21 of the CFR is reserved for rules of the Food and Drug administration. Part 117 contains 7 subparts A to G. Subpart B gives the cGMPs and Subpart C gives hazard analysis and risk based preventive controls. The subpart B addresses the following pre-requisite programs:

117.10 Personnel

117.20 Plant and Grounds

117.35 Sanitary operations

117.37 Sanitary facilities and Controls

117.40 Equipment and Utensils

117.80 Processes and Controls

117.93 Warehousing and Distribution

117.95 Holding and distribution of human food byproducts for use as animal food

117.110 Defect action levels

5.2.4. Pre-requisite Program of the European Union (EU)

In the European Union (EU), the "Hygiene Package" addresses the pre-requisite requirements in 'horizontal' legislation (Regulation 852/2004) and 'vertical' or commodity-specific legislation (Regulation 853/2004 laying down specific hygiene rules for food of animal origin, including fish and fishery products). The hygiene rules in "Section VIII – Fishery Products" of Regulation 853/2004 address the following:

Chapter I	Requirement of vessels
Chapter II	Requirements during and after landing
Chapter III	Requirements for establishments, including vessels, handling fishery products
Chapter IV	Requirements for processed fishery products
Chapter V	Health standards for fishery products
Chapter VI	Wrapping and packaging of fishery products
Chapter VII	Storage of fishery products
Chapter VIII	Transport of fishery products

5.3. Major Requirements of a Pre-requisite Program

The requirements are given below mainly based on the two international recommended Codex codes of practice, with some additional scientific and technical information.

5.3.1. Fishing and Harvesting Vessel Design and Construction

Many types of vessels are used for fishing around the world. The basic requirements are to ease cleaning and to minimize fish damage, contamination and spoilage in vessels, in order to provide high-quality fresh fish. The design and construction of a fishing vessel should consider the following aspects:

For ease of cleaning and disinfection

- The vessels should be designed and constructed to minimize inside corners and projections, to avoid dirt traps
- Construction should facilitate ample drainage
- Should have good supply of clean water or potable water at adequate pressure

To minimize contamination

- All surfaces in handling area should be non-toxic, smooth, impervious and in sound condition to minimize the buildup of fish slime, blood, scales and guts, and to reduce the risk of physical and microbial contamination
- Adequate facilities should be provided for handling and washing fish and shellfish; and adequate supply of cold potable water or clean water
- Adequate facilities should be provided for washing and disinfecting equipment
- Intake of clean water should be easily located to avoid contamination
- Plumbing and waste lines should cope with peak demand

- Non-potable water should be identifiable and separated from potable water to avoid contamination
- Objectionable substances (bilge water, smoke, fuel oil, grease, drainage and solid wastes) should not contaminate the fish and shellfish
- Containers for offal and waste material should be clearly identified, suitably constructed with a lid and made of impervious material
- Separate and adequate facilities should be provided to prevent the contamination of fish and packaging by poisonous or harmful substances such as oil or grease.
- Adequate handwashing and toilet facilities should be available away from the fish and shellfish handling areas
- Entry of birds, insects or other pests, animals and vermin, should be prevented.

To minimize damage to the fish, shellfish and other aquatic invertebrates

- Surfaces in handling areas should have minimum sharp corners and projections
- The design of boxing and shelving storage areas should prevent excessive pressure being exerted on the fish and shellfish
- Chutes and conveyors should be designed to prevent physical damage caused by long drops or crushing
- Fishing gear and its usage should minimize damage and deterioration to the fish and shellfish

To minimize damage during harvesting of farmed and molluscan shellfish

- Seines, nets and traps should be carefully selected to ensure minimum damage during harvesting
- Harvesting areas and all equipment should be designed for rapid and efficient handling without causing mechanical damage, easily cleanable and free from contamination
- Conveying equipment for live products should be constructed with corrosion resistant material that does not transmit toxic substances and cause mechanical injuries to the products
- Live fish should be transported without overcrowding and minimum bruising

- Care should be taken to maintain the factors such as CO_2, O_2, temperature and nitrogenous wastes while transporting and holding fish live.

5.3.2. Facility Design and Construction

The basic requirements for a new processing plant are the identification of a suitable location and infrastructure availability. A plot must be of adequate size, with easy access by road, rail or water. An adequate supply of potable water and energy must be available throughout the year at a reasonable cost. Proper sanitary sewers must be available for waste disposal. Organic wastes from processing area must be removed before discharging waste water into rivers or the sea, after treatment. Solid waste must be handled in a suitable space away from the plant. Contaminants such as smoke, dust, ash, foul odours (e.g. neighbouring fish meal plant) and bacteria are considered as airborne contaminants (e.g. poultry rearing a source of *Salmonella* spp).

Physical surroundings of the plant must be landscaped and provide attractive appearance to the buyer of products. Surroundings should be free from the attract of rodents and birds. Shrubbery should be at least 10 m away from buildings. A grass free strip covered with a layer of gravel or concrete must be near the outer wall of buildings, which allows thorough inspection of walls and control of rodents. Ground immediately in front of doors and entrances should be paved to minimize dust. All areas around the plant and facilities should be well drained to prevent any standing water, where flies and microorganisms. The food processing facility should provide

- Adequate space for equipment, installations, and storage of materials
- Separation of operations, if needed to avoid cross contamination
- Adequate lightning and ventilation
- Protection against pests

External walls, roofs, doors and windows should be water-, insect- and rodent-proof. The seafood processing facility should include a product flow through pattern to prevent contamination, to minimize process delays, and to prevent cross-contamination between finished and raw materials. Fish and shellfish are highly perishable foods and should be handled carefully and chilled without delay. There should not be interruptions and "dead-ends" in the product flow. Time and temperature conditions are extremely important to prevent bacterial growth. The processing facility should facilitate rapid processing and subsequent storage. The design and construction should consider the following points.

For ease of cleaning and disinfection

- The surface of the walls, partitions, and floors should be made of impervious non-toxic materials

- All surfaces that come in contact with fish and shellfish should be corrosion resistant, light colored and smooth impervious material, easily cleanable
- Walls and partitions should have smooth surface up to a height appropriate for operation
- Floors should be constructed to allow adequate drainage
- Ceilings and overhead fixtures should be constructed and finished to minimize the buildup of dirt and condensation, and the shedding of particles
- Windows should be constructed to minimize the buildup of dirt and fitted with removable and cleanable insect proof screens
- Doors should have smooth non-absorbent surfaces
- Joints between floors and walls should be constructed for ease of cleaning

To minimize contamination

- Layout should be designed to minimize cross contamination. Physical or time separation should be there between clean and unclean areas.
- Cool rooms must be separated from hot rooms where cooking, smoking, retorting etc. can take place. Dry rooms must be separated from wet rooms. Separate rooms must be provided for waste material, chemicals (cleaning and disinfection compounds, insecticides, all toxic materials), packaging materials and wood (for smoking).
- All surfaces in the handling areas should be non-toxic, smooth, impervious and in sound condition to minimize the buildup of fish slime, blood, scales and guts and to reduce the risk of physical contamination
- Working surfaces that come in contact with fish and shellfish should be in sound condition, durable and easy to maintain. They should be made of smooth, non-absorbent and non-toxic materials, and inert to fish, shellfish, detergents and disinfectants under normal conditions
- Adequate facilities should be provided for handling and washing of products with adequate supply of cold potable water
- Suitable and adequate facilities should be provided for storage and production of ice.
- Ceiling lights should be covered or suitably protected to prevent contamination by glass or other materials.
- Ventilation should be sufficient to remove excess steam, smoke and objectionable odours. Cross contamination through aerosols should be avoided.

- Adequate facilities should be provided for washing and disinfecting equipment
- Non-potable water lines should be clearly identified and separated from potable water to avoid contamination
- All plumbing and waste lines should cope with peak demands
- Accumulation of solid, semi solid and liquid wastes should be minimized to prevent contamination
- Containers for offal and waste material should be clearly identified and suitable constructed with a fitted lid and made of impervious material
- Separate and adequate facilities should be provided to prevent the contamination by poisonous or harmful substances, dry storage of materials, packaging materials, offal and waste materials
- Hand washing and toilet facilities should be adequate and isolated form the handling area
- Entry of birds, insects or other pests and animals should be prevented
- Water supply lines should be fitted with back-flow devices
- Adequate lighting should be provided in all work areas

5.3.3. Design and Construction of Equipment and Utensils

Utensils and equipment are used for the handling of fish and fish products on a vessel or in a processing facility. They are constantly in contact with fish, shellfish and their products. The contact surfaces such as utensils, knives, tables, cutting boards, boxes, and containers, conveyer belts, gloves, aprons, etc. are designed of a material that it is easily cleanable. All surfaces should be constructed of non-toxic, non-absorbent material that is resistant to the environment, the food and cleaning and disinfecting agents. Food contact materials such as wood, ferrous metals, brass and galvanized metals should be avoided.

The EC (1992) addresses machinery safety and hygiene regulations. The Directive also sets out a certification system wherein machinery is checked for compliance and tagged with an EC mark. Manufacturers are given two years period to bring new machinery into compliance. The seven basic principles for hygienic design agreed by food machinery professionals are:

1. Machinery materials must be designed and constructed so as to clean easily before use.
2. All surfaces and joints must be smooth without ridges or crevices

3. Assemblies must be designed and constructed with welding or continuous bonding with screws, screwheads and rivets, with minimum projections, edges and recesses.
4. Contact surfaces must be readily cleaned and disinfected, and built with easily dismantled parts with curved inner surfaces
5. Liquid derived from foods, cleaning, disinfecting and rinsing fluids must be easily dischargeable from machinery
6. Machinery must be designed and constructed to prevent liquids or insects from entering and accumulating inside
7. Machinery must be designed and constructed such that lubricants do not come in contact with food.

The design and construction of equipment must avoid dead areas where food can get trapped and bacterial growth take place. Dead ends are thermometer pockets, unused pipework, T-pieces, etc. Any equipment must be designed so that the product flow follows "first in first out" principle. The design and construction of equipment and utensils should consider the following p:

For ease of cleaning and disinfection

- Equipment should be durable and movable and capable of being disassembled to allow for maintenance, cleaning, disinfection and monitoring
- Equipment, container, and utensils in contact with fish and shellfish should be designed to provide for adequate drainage and constructed to be easily cleaned, disinfected and maintained to avoid contamination
- Equipment and utensils should be designed and constructed to minimize sharp inside corners and projections and tiny crevices or gaps to avoid dirt traps
- A suitable and adequate supply of cleaning utensils and cleaning agents approved by the official agency should be provided

To minimize contamination

- All surfaces of equipment in handling areas should be non-toxic, smooth, impervious and in sound condition, to minimize the buildup of fish slime, blood, scales, and guts, and to reduce the risk of physical contamination.
- Accumulation of solid, semi-solid and liquid wastes should be minimized to prevent contamination of fish
- Adequate drainage should be provided in storage containers and equipment
- Drainage should not be permitted to contaminate products

To minimize damage

- Surfaces should have a minimum of sharp corners and projections
- Chutes and conveyers should be designer to prevent physical damage caused by long drops or crushing
- Storage equipment. Should be fit for the purpose find not lead to crushing the product.

5.3.4. Hygiene Control Program

Specific measures need to be taken all points, where contamination exist, in order to a safe and wholesome product. The schedules should be implemented to:

- Prevent the buildup of waste and debris
- Protect the fish, shellfish and their products from contamination;
- Dispose of any rejected material in a hygienic manner
- Monitor personal hygiene and health standards
- Monitor the pest control programs
- Monitor the cleaning and disinfecting programs
- Monitor the quality and safety of water and ice supplies.
- The hygiene control program should take into consideration the following points.

A cleaning and disinfection schedule

A cleaning and disinfection schedule should be made to ensure that all parts of the vessel, processing facility and equipment are cleaned appropriately and regularly. Part of this schedule should include a "clean as you go" policy. Pre-cleaning is done prior to cleaning to prepare the area and equipment for cleaning, which involves steps such as removal of all fish, shellfish and their products from the area, protection of sensitive components and packaging materials from water, removal by hand or squeegee of fish scraps, etc.

A typical cleaning and disinfecting process involve seven separate steps:

1. Pre-rinse: A rinse with water removes the remaining large pieces of loose soil.
2. Cleaning: Cleaning removes soil, food residues, dirt, grease or other objectionable matter.
3. Rinsing: Rinse with potable water or clean water removes all soil and detergent residues.

4. Disinfection: Chemicals (approved by the official agency) and/or heat destroys most micro-organisms on surfaces.
5. Post-rinse: A final rinse with potable water or clean water removes all disinfectant residues.
6. Storage: Storage of cleaned and disinfected equipment, container and utensils to prevent contamination.
7. Checking the efficiency of cleaning: Control the efficiency of the cleaning as appropriate.

Most detergents or cleaning agents work faster and more effectively at higher temperatures. Cleaning is often carried out at 60–80°C in areas where such high temperatures can be used.

The ideal detergent is characterized by the following properties

- It possesses sufficient chemical power to dissolve the material.
- It has a surface tension low enough to penetrate into cracks and crevices; it should be able to disperse the loosened debris and hold it in suspension.
- If used with hard water, it should possess water-softening and calcium-salt-dissolving properties to prevent precipitation and buildup of scale on surfaces.
- It rinses freely from the surfaces, leaving them clean and free from residues, which could harm the products and affect sterilization negatively.
- It does not cause corrosion or other deterioration of surfaces. It is recommended always to check by consulting the supplier of machines, etc.
- It is not hazardous for the operator.
- It is compatible with the cleaning procedure being used, whether manual or mechanical.
- If solid, it should be easily soluble in water and its concentration easily checked.
- It complies with legal requirements concerning safety and health as well as biodegradability.
- It is reasonably economical to use.

A detergent with all these characteristics does not exist. So, choose a usable cleaning agent and water treatment additives to get combined detergent with most of the properties. All cleaning methods require sufficient contact time to fully loosen and suspend soils. A moderately alkaline detergent is normally used in fish processing plants require 10–15 min to fully loosen most processing soils.

Disinfection is affected by physical treatments such as heat, UV irradiation, or by means of chemical compounds. The use of heat in the form of steam or hot water is a very safe and widely used method of disinfection. The most commonly used chemicals for disinfection are given in Table 5.1.

1. Chlorine

Chlorine is one of the most effective and widely used disinfectants. It is available in several forms such as sodium hypochlorite, chloramines and other chlorine containing organic compounds. Gaseous chlorine and chlorine dioxide are also available. Chlorinated disinfectants at 200 ppm free chlorine are very active. The effect gets reduced when organic residues are present. The chlorine dissolved in water produces hypochlorous acid (HOCl), which act by means of oxidation as disinfectants. It is highly unstable in solution and it liberates as toxic chlorine gas in acid solution. It is also more corrosive at low pH. However, the germicidal activity is better in acid than in alkaline solution. Organic chlorinated disinfectants are more stable but require longer contact times. At 200 ppm free chlorine level, chlorinated disinfectants are non-corrosive to high quality stainless steel, but are corrosive to other less resistant materials.

Table 5.1. Most Commonly used Disinfectants in Fish Processing Plants

Disinfectant	**Forms/ Description**
Chlorine	Hypochlorite, Chlorine gas, Organic chorine e.g., Chloramines
Iodophors	Iodine dissolved in surfactant and acid
Quaternary Ammonium Compounds (QAC)	Benzalkonium chloride and related compounds (Quats)
Acid-Anionic	Combination of certain surfactants and acids
Peroxy Compounds	Acetic acid and hydrogen peroxide (peroxyacetic acid)
Carboxylic Acid	Fatty acids combined with other acids (fatty acid sanitizers)
Chlorine Dioxide	A gas formed onsite and dissolved in solution or by acidification of chlorite and chlorate salts
Ozone	A gas formed onsite and dissolved in solution
Hot Water/Heated Solutions	Water at 77-88°C

Chlorine is cheap and widely available and hence, monitoring the free residual level is simple. WHO (1996) recommends 5 mg chlorine/litre (5 ppm) for disinfection. There should be a residual concentration of free chlorine 0.5 mg/l (0.5 ppm) after at least 30 min contact time at pH <8.0. For disinfection of clean equipment up to 200 mg/l (200 ppm) is used. To avoid corrosion a lower concentration of 50-100 ppm and longer contact times of 10-20 min are recommended.

2. Iodophors

Iodophors contain iodine bound to a non-ionic carrier, which releases iodine. They have broad antimicrobial spectrum like chlorine. The effect is maximum at 25 ppm free iodine concentration and at pH 2-4. The effect reduces in the presence of organic material. Commercial formulations are acidic, which helps to dissolve scales. They are corrosive. If used at temperature above 45°C, free iodine liberates out. If residues of product and caustic cleaning agents are present, they combine with iodophors and cause very unpleasant "phenolic" off-flavors.

3. Hydrogen peroxide and peracetic acid

They are effective disinfectants act by means of oxidation with a broad antimicrobial spectrum. They are used as dilute solutions alone or in combination for other disinfectants. The presence of organic material inactivates them. The effective concentration is 200-300 ppm.

4. Quaternary ammonium compounds

They are cationic surfactants more effective against fungi and bacteria, but less effective against Gram negative bacteria. They are used as an alternative to other disinfectants to avoid development of resistant strains of microorganisms. They have good penetrating properties due to low surface tension, but difficult to rinse off. If they come into contact with anion-active detergents, they precipitate out and become inactivated. The effective concentration is 200 ppm on food contact surfaces. Table 5.2 gives the concentrations of commonly used disinfectants.

Table 5.2. Disinfectant concentrations commonly used in Fish Processing Plants

Disinfectant	Food contact surface	Non-food contact surfaces	Plant water
Chlorine	100-200 ppm	400 ppm	3-10 ppm
Iodine	25 ppm	25 ppm	-
Quats	200 ppm	400-800 ppm	-
Chlorine dioxide	100-200 ppm	100-200 ppm	1-3 ppm
Peroxyacetic acid	200-315 ppm	200-315 ppm	

5. Other disinfectants

Chloramines are more stable but less microbiocidal, and less efficient in killing parasites and virus than chlorine. **Chlorine dioxide** is more microbiocidal than chlorine at high pH, but by-products are a major concern. **Ozone** and **UV** leave no residual matter to monitor. Ozone is very efficient in killing protozoa. The efficiency of UV disinfection decreases if there is any turbidity or dispersed organic matter is present.

The resistance of the microorganisms to most disinfectants follow this order of sensitivity: vegetative bacteria > viruses > bacterial spores, acid-fast bacteria and protozoan cysts. Indicator bacteria are more sensitive microorganisms and therefore, **fecal coliforms** give a clear indication on the potentially pathogenic microorganisms in water.

Designation of personnel for cleaning

- Handlers or cleaning personnel should be well trained in the use of special cleaning tools and chemicals, and in the methods of dismantling equipment for cleaning.
- In each processing plant or vessel, a trained individual should be designated to be responsible for the sanitation of the processing facility or vessel and the equipment therein.

Maintenance of premises, equipment and utensils

- Buildings, materials, utensils and all equipment in the establishment including drainage systems should be maintained in a good state and order.
- Equipment, utensils and other physical facilities of the plant or vessel should be kept clean and in good repair.
- Procedures for the maintenance, repair, adjustment and calibration of apparatus should be established.

5.3.5. Pest Control Program

Pests such as rodents, birds, insects, dogs and cats can carry a variety of human disease agents and introduce them into the processing environment. The presence of pests in a processing plant is unacceptable and a pest control program should be based on three principles:

GHP should be employed to avoid creating an environment conducive to pests

PCP should include preventing access, eliminating harbourage and infestations, and establishing monitoring detection and eradication systems

Physical, chemical and biological agents should be properly applied by appropriately qualified personnel.

5.3.6. Supply of Water, Ice and Steam

1. Water

- An ample supply of cold and hot potable water and/or clean water under adequate pressure should be provided.
- Potable water should be used wherever necessary to avoid contamination.

Quality supply of water is very important for cleaning and processing. WHO (1993) published extensive guidelines on drinking water quality. The microbiological criteria suggested are given in Table 5.3.

Table 5.3. Bacteriological quality of drinking water

Organisms	**Guideline value**
All water intended for drinking	
E. coli or thermotolerant coliform bacteria	Not detectable in 100-ml sample
Treated water entering the distribution system	
E. coli or thermotolerant coliform bacteria	Not detectable in 100-ml sample
Total coliform bacteria	Not detectable in 100-ml sample
Treated water in the distribution system	
E. coli or thermotolerant coliform bacteria	Not detectable in 100-ml sample
Total coliform bacteria	Not detectable in 100-ml sample. In the case of large supplies, where sufficient samples are examined: Not detectable in 95% of samples taken during any 12-months period

2. Ice

- Ice should be manufactured using potable water or clean water.
- Ice should be protected from contamination.

3. Steam

- An adequate supply of water at sufficient pressure should be maintained for operations that require steam.
- Steam used in direct contact with fish or shellfish or food contact surfaces should not constitute a threat to the safety or suitability of the food.

5.3.7. Waste Management

Offal and other waste materials from the processing area and factory premises must be removed on a regular basis. Separate facilities for containment of offal and waste material must be available. A hygienic waste water disposal system must be properly maintained. Sewage disposal must be done into an adequate sewerage system or disposed of through other proper means.

5.3.8. Personnel Hygiene and Health

Personnel hygiene and facilities should ensure that an appropriate degree of personnel hygiene is maintained to avoid contamination.

Facilities and equipment

- Adequate means for hygienically washing and drying hands.
- Adequate toilet and changing facilities for personnel should be suitably located and designated.

Personnel hygiene

- No person suffering from, or carrier of any communicable disease or an infected wound or open lesion should be engaged in the preparation, handling or transportation of fish.
- Adequate and appropriate protective clothing, head covering and footwear should be worn.
- All persons should maintain a high degree of personal cleanliness and should take all necessary precautions to prevent contamination of products or facilities.
- Hand-washing should be carried out by all personnel working in a processing area at the start of fish or shellfish handling activities and upon re-entering a processing area immediately after using the toilet.

The following should not be permitted in handling and processing areas:

- Smoking, spitting, chewing or eating, sneezing or coughing over unprotected food.
- The adornment of personal effects such as jewelry, watches, pins or other items that, if dislodged, may pose a threat to the safety and suitability of the products should not be allowed.

5.3.9. Transportation

Vehicles should be designed and constructed such that:

- Walls, floors and ceilings, where appropriate, are made of a suitable corrosion-resistant material with smooth non-absorbent surfaces. Floors should be adequately drained.
- Chilling equipment should be used to maintain chilled fish or shellfish during transport to a temperature as close as possible to 0 °C or, for frozen fish, shellfish and their products, to maintain a temperature of –18 °C or colder (except for brine frozen fish intended for canning, which may be transported at –9 °C or colder).
- Live fish and shellfish are to be transported at temperatures that the species can tolerate.

- Fish or shellfish should be provided protection against contamination, exposure to extreme temperatures and the drying effects of the sun or wind.
- Mechanical refrigeration used should permit the free flow of chilled air around the load.

5.3.10. Product Tracing and Recall Procedures

A system for recall of product is a necessary component of a pre-requisite program. Product tracing (lot identification) is essential to an effective recall procedure.

- Managers should ensure effective procedures are in place to have complete product tracing and rapid recall of any lot of fishery product from the market.
- Appropriate records of processing, production and distribution should be kept and retained for a period that exceeds the shelf-life of the product.
- Each container of fish, shellfish and their products intended for the final consumer or for further processing should be clearly marked to ensure the identification of the producer and of the lot.
- Where there is a health hazard, products produced under similar conditions, and likely to present a similar hazard to public health, may be withdrawn. The need for public warnings should be considered.
- Recalled products should be held under supervision until they are destroyed, used for purposes other than human consumption, or reprocessed in a manner to ensure their safety.

5.3.11. Training

Fish or shellfish hygiene training is of fundamental importance. All personnel should be aware of their role and responsibility in protecting fish or shellfish from contamination and deterioration. Handlers should have the necessary knowledge and skill to enable them to handle fish or shellfish hygienically. Personnel who handle strong cleaning chemicals or other potentially hazardous chemicals should be instructed in safe handling techniques.

Each fish and shellfish facility should ensure that individuals have received adequate and appropriate training in the design and proper application of an HACCP system and process control. Training of personnel in the use of HACCP is fundamental to the successful implementation and delivery of the program in fish or shellfish processing establishments. Managers should also arrange for adequate and periodic training of relevant employees in the facility to understand the principles of HACCP.

References

CAC 2000. Proposed Draft. Code of Practice for Fish and Fishery Products. Alinorm 01/18. Codex Alimentarius Commission of Food and Agriculture Organization / World Health Organization, Rome, Italy.

CAC 2001. Food Hygiene Basic texts. 2nd ed. Codex Alimentarius Commission of Food and Agriculture Organization / World Health Organization, Rome, Italy.

EC (1993. Council Directive 93/43/EEC of 14 June 1993 on the hygiene of foodstuffs. Official Journal of the European Commission L 175, 19/07/1993 pp. 0001-0011.

EC 1991. Council Directive 91/493/EEC of 22 July 1991 laying down the health conditions for the production and the placing on the market of fishery products. Official Journal of the European Commission L 268, 24/09/1991 pp. 0015 - 0034.

EC 1992. Proposal for a council directive on the hygiene of foodstuffs. Official Journal of the European Commission C 24/11, 31/01/1992 pp. 11-16.

EC 1998. Council directive 98/83/EC of 3 November 1998 on the quality of water intended for human consumption. Official Journal of the European Commission L330, 05/12/1998. pp 0032-0054.

FDA 1995. Procedures for the Safe and Sanitary Processing and Importing of Fish and Fishery Products; Final Rule. Code of Federal Regulations, Parts 123 and 1240. Volume 60, No 242, 65095-65202. Us Food and Drug Administration

FDA 2001. Current Good Manufacturing Practices. 21 CFR Part 110. http://seafood.ucdavis.edu/GUIDELINES/gmps.htm Us Food & Drug Administration.

Huss HH and Ryder J. 2003. Pre-requisite to HACCP. Chapter 7. In: Huss HH, Ababouch L, Gram L. (eds). Assessment and management of seafood safety and quality, FAO Fisheries Technical Paper. No. 444. Rome, FAO. 230p.

ICMSF 1988. Microorganisms in Foods 4. Application of the Hazard Analysis Critical Control Point (HACCP) system to ensure microbiological safety and quality. International Commission on Microbiological Specifications forFoods, Blackwell Scientific Publications.

NACMCF 1998. Hazard analysis and critical control point principles and application guidelines. Journal of Food Protection 61, 762-775, National Advisory Committee on Microbiological Criteriafor Foods.

Ryder, J., Karunasagar, I. and Ababouch L., eds. 2014. Assessment and management of seafood safety and quality: current practices and emerging issues. FAO Fisheries and Aquaculture Technical Paper No. 574. Rome, FAO. 432 pp.

WHO 1993. Guidelines for drinking water quality. 2nd ed. Vol 1. World Health Organization, Geneva, Switzerland.

WHO 1999. Strategies for implementing HACCP in small and/or less developed businesses. Report on WHO Consultation. WHO/SDE/PHE/FOS/99.7. World Health Organization, Geneva, Switzerland.

6

Application of HACCP System for Seafood Industries

Application of HACCP concept in surveillance & quality assurance programs for raw, frozen, canned, cured, irradiated, cooked, chilled, MAP & freeze-dried products.

HACCP

Hazard analysis and critical control point (HACCP) is a system that identify, evaluate and control hazards that are significant for food safety (CAC, 2003). It is a science based systematic approach that assesses hazards and establishes control systems to focus on prevention rather than end product testing. It enhances the safety of the product from biological, chemical, and physical hazards in a production process. It provides way to demonstrate competence to consumers and compliance with legislative requirements.

6.1. History

HACCP was originally developed by Pillsbury Company in 1959 to produce zero defect food for astronauts of National Aeronautics Space Administration (NASA) in USA. In 1971, the HACCP concept was presented at the United States National Conference on Food Protection. The first major document on HACCP was approved by the National Advisory Committee on Microbiological Criteria for Foods (NACMCF) in 1989. The Codex then issued the first HACCP guidelines in 1993, which was adopted by the FAO/WHO Codex Alimentarius Commission (CAC). The USFDA's Seafood HAACP program became mandatory in USA since 1997. The European Union, in 2004, laid down the general hygiene requirements of all food business operators (EC Regulation 852/2004) and the additional specific requirements for foods of animal origin, including live bivalve mollusks and fishery products which (Regulation EC 853/2004). The United States Food Safety Modernization Act 2011 became the latest enforcement of food safety measures to protect public health. Thus, HACCP has become an International reference system for food safety assurance.

6.2. Application of HACCP System

The FDA and the food canning industries in USA are the first to adopt the preventive controls and HACCP concepts. In mid 1980s, HACCP became a major focus of regulatory agencies and industries in all countries, thus is termed as an industry driven program. Regulatory agencies certify the food facilities by conducting on-site verification on proper implementation of HACCP. Currently, most of the national food control agencies, including FSSAI (India) have adopted the HACCP system. The FDA "Fish and Fishery Products Hazards and Controls Guidance" is a specific reference guide for fish and fishery products. The HACCP system is applied from fish production to fish consumption.

6.2.1. Basic Principles of HACCP

The CAC adopted the basic principles of food hygiene, including HACCP and revised the guidelines in 2003 (CAC, 2003). The HACCP system consists of seven principles.

Principle 1: Conduct a hazard analysis

Principle 2: Determine the critical control points (CCPs)

Principle 3: Establish critical limits

Principle 4: Establish monitoring procedures

Principle 5: Establish corrective actions

Principle 6: Establish verification procedures

Principle 7: Establish record-keeping and documentation procedures.

6.2.1.1. Developing a HACCP Plan

Prior to the application of HACCP to any seafood establishment, a proper prerequisite program should be implemented, as recommended by International Code of Practice – General Principles of Food Hygiene (CAC, 2003) Annex: HACCP system and Guidelines for its Application and in accordance to the Code of Practice for Fish and Fishery Products (CAC, 2008). The conditions and requirements of prerequisite program are discussed in the previous lesson. The management of the establishment is responsible for the implementation of an effective HACCP system. The effectiveness depends mainly on the knowledge and skill of the employees. If required, an external expert advice shall be taken from the trade and industry associations, independent experts, or regulatory authorities. Prior to the application of the HACCP principles in a HACCP plan, five preliminary tasks have to be fulfilled (CAC, 2003).

6.2.1.1.1. Assembling the HACCP Team

The first task is to assemble a HACCP team to develop a HACCP plan. Team should have expertise in food safety and quality, food technology and quality assurance. Outside experts can be included, if expertise is not available within the establishment. The HACCP team must have access to all relevant and important information to identify the hazards and the control measures. The HACCP team shall be constituted by safety and quality supervisor, technical supervisor, equipment maintenance supervisor, key personnel of operation, and an advisor.

6.2.1.1.2. Describe the Food and Its Distribution

The HACCP team describes the product, including safety information. The description should include harvesting area and technique, raw materials and ingredients, factors influencing safety such as composition, physical and chemical parameters (water activity, salt, pH), processing methods (heating, freezing, brining or smoking), packaging, storage conditions, distribution methods, and shelf life under specified conditions.

***Example**: The product, canned tuna in olive oil is destined for export to Europe and the USA. It is generally consumed without any cooking, as an appetizer, in a sandwich or salads. It is consumed by the public with no specific age restriction.*

6.2.1.1.3. Describe the Intended Use and Consumers of the Food

The intended use depends on the end user or consumer, as the method of preparation before the consumption greatly influence the safety of the product. The HACCP team must also describe the expected consumers of the food i.e. whether the intended consumers are general public or a particular segment of the population like infants, the elderly, etc.

***Example:** Live oysters (Crassostrea gigas) harvested from Tuticorin, depurated for at least 44 hours, using UV disinfected water. The depurated oysters are packed in mesh nets and sold live to retailers and to restaurants.*

6.2.1.1.4. Develop a Flow Diagram which Describes the Process

The HACCP team develops a flow diagram to provide a clear, simple outline of all steps involved in the process, before and after the processing. Time and temperature conditions during processing should be mentioned A block type diagram is sufficient. A simple schematic of the facility is useful to understand the product or process flow.

6.2.1.1.5. Verify the Flow Diagram

The HACCP team conducts an on-site review of the operation, to verify the accuracy and completeness of the flow diagram and documents the modifications, if any.

6.2.1.2. Principles of HACCP

The seven principles of HACCP are applied after the completion of five preliminary tasks.

6.2.1.2.1. Conduct of Hazard Analysis

A hazard is defined as a biological, chemical or physical agent in or condition of food such as temperature abuse, insufficient thermal process, having the potential to cause an adverse effect and harm to consumers. The HACCP team should list out all hazards expected to occur in production, processing, transportation and distribution until the food reaches the consumer. An accurate hazard analysis is essential. The HACCP team should also identify which hazards can be eliminated or reduced to acceptable levels in the production. The aspects to be considered while conducting hazard analysis are:

- Whether the raw materials and ingredients contain any hazardous agents?
- Will the intrinsic factors permit survival, multiplication of pathogen or toxin formation in seafood?
- Whether the processing conditions reduce or destroy contaminants or pathogens ? Are there any possibilities for recontamination?
- Does the packaging affect the microbial population?
- Will the food be heated or cooked before consumption?
- Is the product destined for the general public or for consumption by the population with high susceptibility of illness such as infants, elderly persons or patients?

A decision tree with a number of questions can be used to the "real" potential hazards at each step of the processing chain and for all hazards. An element of risk assessment is involved in the evaluation of potential hazards. Only those hazards likely to occur and cause a reasonable adverse health effect are regarded as significant hazards.

The HACCP team must also consider the application of already existing control measure for each hazard. More than one control measure is required to control a specific hazard (s) and more than one hazard is also controlled by a specific control measure. The hazards associated with each step in the production should be listed along with any control measures. A hazard worksheet is used to organize and document the food safety hazards (Table 6.1)

Table 6.1. Hazard Analysis Worksheet

Company Name: **Company Address:**			**Product Description :** **Method of Storage and Distribution:** **Intended use and Customer:**		
(1)	**(2)**	**(3)**	**(4)**	**(5)**	**(6)**
Ingredients/ processing step	Identify potential hazards introduced, controlled or enhanced at this step	Are any potential of safety hazards significant ? (Yes/No)	Justify your decision for Column 3	What preventive measures can be applied to prevent the significant hazards?	Is this step a critical control point ? (Yes/No)
Step 1	Biological				
	Chemical				
	Physical				

6.2.1.2.2. Determine Critical Control Points (CCPs)

A CCP is a point, step or procedure at which control can be applied. CCP is essential to prevent or eliminate or reduce a food-safety hazard to an acceptable level. A single hazard can be controlled by more than one CCP. Likewise, more than one hazard can be controlled at a single CCP. The team should identify all CCPs accurately to control food safety hazards, which can also be facilitated by a decision tree, as recommended by Codex (CAC, 2003) (Fig 6.1)

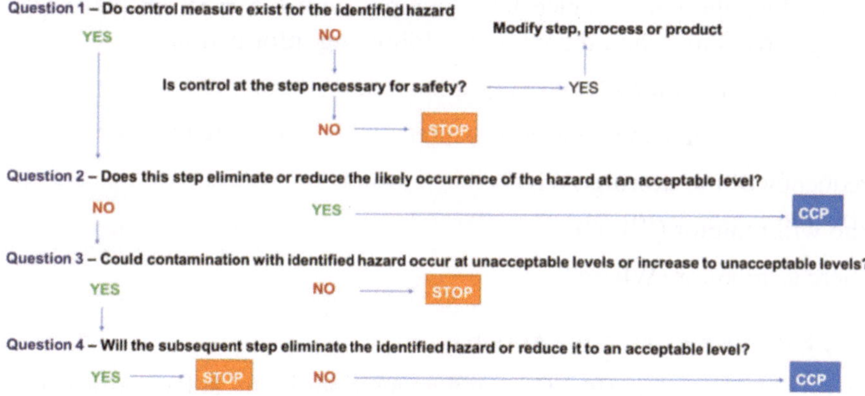

Fig 6.1. Decision tree for the identification of CCP

If a hazard is identified at a step and if no control measure exists at that step, then the product or process should be modified at that step or at an earlier or a later step to include a control measure. This process must be done at each step and for each hazard to identify all CCPs.

6.2.1.2.3. Establish Critical Limits

A critical limit is a maximum and/or minimum value at which a biological, chemical or physical parameter is controlled at a CCP. It is a boundary to judge whether an operation is producing safe products or not. Each CCP can have one or more control measures. Each control measure can have one or more critical limits. The critical limits are scientifically based on measurable parameters such as temperature, time, chlorine levels, humidity, moisture level, water activity, pH, titratable acidity, salt concentration, available chlorine, viscosity, preservatives or sensory quality. If these parameters are maintained within the limits, the hazard is under control at a CCP. As far as possible, avoid microbiological limits. If required, rapid techniques can be used.

The critical limits should preferably meet the requirements of government regulations and / or company standards supported by scientific data. Authoritative critical limit information is available at sources such as Fish and Fishery Products Hazards Control Guidance (FDA, 2011), and other scientific publications or documents.

6.2.1.2.4. Establish Monitoring Procedures

Monitoring is a planned sequence of observations or measurements to assess whether a CCP is under control. The monitoring procedures can detect the loss of control at a CCP and ensure whether critical limits are exceeded. The purpose of monitoring is to measure the performance level of system's operation at the CCP (trend analysis), when there is a loss of control at a CCP, and to establish records whether the performance level comply with the HACCP plan. The monitoring procedures should contain the following information

1. What will be monitored? (What?)
2. How critical limits and control measures will be monitored (How?)
3. Frequency of monitoring (Why?)
4. Who will monitor (Who?)
5. Where to monitor (Where?)

6.2.1.2.5. Establish Corrective Actions

The main objective for HACCP implementation is to solve or prevent problems. Corrective actions must be pre-defined and taken when the results of monitoring at a CCP indicate a loss of control. The establishment must have procedures to identify, separate, mark and control the products that indicate a loss of control. The corrective actions should determine the cause of the problem, define the act to prevent recurrence, and follow up with monitoring and reassessment. The corrective actions taken on the control and disposition of the product must be recorded, which helps in verification.

6.2.1.2.6. Establish Verification Procedures

Verification is done to determine whether the HACCP system is working effectively in compliance with HACCP plan, because mere implementation does not guarantee the effectiveness. Verification applies methods and procedures such as random sampling, analysis and monitoring. Verification is done by an unbiased internal or external qualified individual (s) and the activities are documented in HACCP plan. Verification must take place when there is a change in raw materials, product formulation, processing procedures, consumer and handling practice, new information on hazards, consumer complaints, recurring deviation or any other indication.

6.2.1.2.7. Establish Record Keeping and Documentation

Records and documentation are essential to review the adequacy and adherence of HACCP system to the HACCP plan. The records relevant to the HACCP program are:

1. Support documents for developing the HACCP plan (including validation records
2. Records generated by the HACCP system (including monitoring records)
3. Deviation and correction action records, verification/validation records
4. Documentation on methods and procedures
5. Records of employee training programs

The records are made in different forms such as processing charts, written procedures and tables; and stored in paper or electronic forms with integrity.

6.3. HACCP for Seafood Industries

HACCP system is implemented by all wholesale processors and distributors of seafood. All seafood manufacturers adhere to current GMPs, seafood HACCP and all applicable regulations. Retail establishments, fishing vessel, and transporters who do not process seafood and practices for storage on-board; and therefore, are not required to operate under a HACCP system. The hazards associated with the harvest and transportation of such seafood are evaluated by the receiver.

The FDA has established Preventive Control type Regulation (21 CFR Part 123) for Fish and Fishery products based on HACCP concept. The seafood HACCP regulation requires seafood processors to identify food safety hazards and to develop plans for the control of those hazards. In addition, seafood HACCP regulation requires importers to comply with requirements designed to ensure that imported products are processed in accordance with the seafood HACCP regulation. Some seafood products are subject to 21 CFR Part 113 applicable to thermally processed low acid foods packaged in hermetically sealed containers. 21 CFR Part 117 provides current Good Manufacturing Practice, Hazard Analysis, and Risk based Preventive Controls for Human Foods (cGMP and PC Regulation) and 21 CFR part 121 gives Mitigation Strategies to Protect Food against Intentional Adulteration (IA regulation).

In European Union, the legislation (EC Regulation No.852/2004 and 853/2004) on hygiene regulations requires food business operators, except primary producers should implement and maintain procedures, based on the principles of HACCP.

6.3.1. Application of HACCP Concept for raw, and Processed Fish Products

1. Seafood are classified into 9 major groups
2. Molluscan shellfish
3. Raw fish to be eaten without any cooking
4. Fresh or frozen fish and crustaceans fully cooked before consumption.
5. Lightly preserved fish products i.e. NaCl <6% in water phase, pH >5.0. The prescribed storage temperature is <5°C. e.g. salted, marinated, cold smoked and gravid fish
6. Fermented fish, i.e. NaCl <8%, pH neutral to acid. Storage at ambient temperature
7. Semi-preserved fish i.e. NaCl >6%, or pH < 5, preservatives (sorbate, benzoate, nitrite) may be present. The prescribed storage temperature is <10°C. e.g. salted and/or marinated fish or caviar, fermented fish
8. Mildly heat-processed (pasteurized, cooked, hot smoked) fish and crustaceans, including pre-cooked, breaded fillets. The prescribed storage temperature is <5°C
9. Heat-processed (sterilized, packed in sealed containers)
10. Dried, smoke-dried fish, heavily salted fish. Storage at ambient temperatures.

6.3.1.1. Hazard Analysis of Raw Material

A large number of hazards are related to the pre-harvest situation or raw material handling. Most fish and shell fish are caught from wild population and there are specific safety hazards associated with wild fish. The intensive aquaculture fish have new and increased risks. In a hazard analysis of the pre-harvest conditions, the significant hazards are identified are given in Table 6.2.

Table 6.2. Hazard analysis of pre-harvest conditions and raw material handling

Hazards	**Potential hazard**		**Analysis of hazard**			**Control**		
	Contamination	**Growth**	**Severity**	**Likely occurrence**	**Significant**	**Govt. monitoring program**	**Prerequisite program**	**Inclusion in HACCP plan**
Indigenous pathogenic bacteria	-	+	high	high	+	-	-	+
Non-indigenous pathogenic bacteria	+	+	high	high	+	+	+	+
Viruses	+	-	high	High/low	+/-	+	+	+
Biotoxins	+	-	high	High/low	+/-	+	-	+
Biogenic amines	-	+	low	High/low	+/-	-	-	+
Parasites	+	-	low	high	+	-	-	+
Chemicals	+	-	medium	High/low	+/-	+	-	+

Significant hazards along with preventive measures are

1. **Pathogenic bacteria**: Pathogenic bacteria are present in low numbers in all fish and shellfish at the time of harvest and unlikely to cause disease. When time/temperature abuse occurs, growth and toxin production take place reaching to unsafe levels, which then becomes a significant hazard in fish to be eaten raw or in product that is not heat treated. High number of pathogenic *Vibrio* spp accumulate in bivalves. Pathogenic bacteria from animal/human reservoir can also present a significant hazard in fish to be eaten raw due to low infective dose.

   Preventive measures

   a. Control and monitoring of harvest areas for fecal pollution

   b. Limit the time between harvest and refrigeration.

2. **Viruses :** Presence of viruses is a hazard in molluscan shellfish because of contamination from sewage, filter feeding nature, and raw consumption

   Preventive measure:

   a. Control and monitoring of harvest areas for fecal pollution.

3. **Biotoxins :** Natural toxins accumulate in fish that feed on marine algae form a hazard. Fish from the tropical and subtropical areas (ciguatera) and in shellfish worldwide are more prone.

   Preventive measures

   a. Harvest shellfish from safe waters. All shellfish containers must bear a tag that identifies the type and quantity of shellfish, the harvester, location and date of harvest.

   b. Ensure fish is harvested from an area free from CFP problem

4. **Biogenic amines:** Amines are produced by certain fish species, when there is time/temperature abuse and can cause illness. It is a post-harvest hazard, but considered as pre-receiving hazard introduced during handling and transportation.

   Preventive measures

   a. Rapid chilling of fish immediately after capture. Pack the fish in ice or chilled seawater within 12 h after catch. In case of large fish (tuna), chill to an internal temperature of 10°C or less within 6 h after capture.

5. **Parasites:** Parasites are present in significant numbers in certain wild-caught and aquaculture fish, if they feed processing wastes or bycatch fish.

   Preventive measure

   a. Identify parasites during processing of fish to eliminate them

6. **Chemicals:** Fish harvested from freshwater, estuaries and near shore waters and from aquaculture contain unsafe levels of chemicals, which represent a significant hazard.

   Preventive measures:

   a. Check for the presence of Govt controlled monitoring programs

   b. Ensure fish are not harvested from waters closed for commercial fishing

   c. Full control of chemical contamination of the environment (soil/water) surrounding aquaculture site

   d. Control of water quality

   e. Control of the feed supply

   f. Use of approved agrochemicals and drugs as per manufacturer's instructions.

   g. Correct withdrawal of veterinary drugs

Seafood processors have no control and only limited information on the history of the raw material. So, there is a need to identify the significant hazards associated with the raw material and devise control measures before it is received in the company premises. The **receiving step** is the first CCP in any seafood processing, and the monitoring procedures are mainly to check documents relating to certificates of origin, harvester, date and location of harvesting, copies and results of government monitoring programs, etc.

6.3.1.2. Hazard Analysis of Molluscan Shellfish

Molluscan shellfish are harvested either by raking or trawling from the bottom (oysters and mussels) or by digging from the sand at low tide (clams and cockles). They are sorted, washed and packed in bags or crates or left in a pile on deck, transported, and sold live to the consumer or processor. The processors apply heat to facilitate shucking to relax the adductor muscle, which are then washed, packed and sold fresh, frozen or further processed and canned. Most mollusks (oysters, mussels, clams and cockles) that grow in shallow, near-shore estuarine waters are likely to get contaminated with sewage-derived pathogens (pathogenic bacteria, and viruses) and from the environment (biotoxins and chemicals).

Significant hazards identified are

a. Contamination with pathogens, biotoxins, and chemicals from the harvesting area
b. Contamination of pathogens during processing
c. Growth of pathogens during processing and storage.

Preventive measures are

a. Control and monitoring of harvesting area. Check for tags and licensed harvesters or certified dealers. Depuration. A warning label to inform consumers of the risk
b. Control of contamination during processing by prerequisite program
c. Limit the time from harvest to refrigeration. Proper chilling ($<5^oC$) at all times of storage. Control of temperature is part of prerequisite program.

The CCPs identified are

a. Receiving step
b. Labelling step

Critical limits

a. Check for tags in all containers that discloses the date and place of harvest, quantity, name and license number of harvester. No entry for mollusks from closed areas.

Monitoring program

a. What : Tags, labels, license of fisher

b. How : Visual check

c. When : All containers

d. Who : Receiving employee, supervisor or quality control (QC) staff

Corrective actions, Record-keeping and Verification

a. Reject, if untagged or from closed areas

b. Receiving records of all shellfish

c. Daily review of records

6.3.1.3. Hazard Analysis of Raw Fish to be EatenRaw

Th hazards identified in raw products are associated with the pre-harvest / pre-receiving conditions.

Significant hazards identified are

a. Contamination of fish with non-indigenous bacteria, viruses, biotoxins or environmental chemical contaminants (heavy metals, pesticides, drugs in aquaculture)

b. Presence of parasites

Preventive measures are

a. Control and monitoring of harvesting areas. Control of the use of drugs in aquaculture

b. Control of contamination (bacteria, viruses) during processing by pre-requisite program

c. Prohibition of the use of puffer fish for human consumption

d. Avoidance of fish with a record of ciguatera

e. Introduction of a freezing step to eliminate the risk from parasites.

The CCPs identified are

a. Receiving step

b. Freezing step

Critical limits

a. Check for a source control or a certificate in all lots of fish.

b. Ensure fish is not harvested from waters closed to fishing.

c. Ensure aquaculture fish is not contaminated with drug.

Monitoring program

a. What: Time and temperature at freezing step; Tags, labels, & license of fisherman

b. How: Visual check

c. When: All containers; Continuous recording of freezing temperature

d. Who: Receiving employee, Supervisor or QC-staff

Corrective actions

a. Reject, if fish is untagged or caught from closed areas

b. Adjust freezer or refreeze material, if not properly frozen

Record keeping and Verification

a. Receiving records on all fish raw material (quantity, harvesting details)

b. Temperature records

c. Daily review of records

6.3.1.4. HACCP Analysis of Fresh/Frozen Fish and Crustaceans

Fish and crustaceans caught in the sea or freshwater are handled and processed without any use of additives or chemical preservatives and distributed with chilling or freezing.

Significant hazards identified are

a. Presence of biotoxins: Applies to fish from warm waters with ciguatera history and to puffer fish

b. Formation of histamine: Applies mostly to scombroid fishes

c. Presence of chemicals: Applies only to fish from aquaculture or coastal areas.

For all other fish, there are no safety hazards and no HACCP plan are required.

Preventive measures are

a. Sorting of the catch to exclude puffer fish and fish with CFP advisory or problem.

b. Rapid chilling of fish immediately to <10°C. Chilling towards the freezing point is desirable. Control of temperature is part of the prerequisite program

c. Collect information on capture area with government ban on fishing.

The CCPs identified is

a. Receiving step

Critical limits

a. No puffer fish or fish with CFP advisory or problem is allowed in processing.
b. No fish harvested in an area closed for fishing is allowed in processing
c. The critical limit for histamine is <50 ppm

Monitoring program

a. What: Sorting procedures, tags, labels, harvesting vessels, record decomposition of lot. Temperature records
b. How: Visual check
c. When: All lots
d. Who: Receiving employee

Corrective actions

a. Reject lots with no information on catching area, or if from closed area
b. Reject the lot or perform histamine analysis on lots of poor sensory quality
c. Inform harvester, adjust cooling procedures

Record keeping and Verification

a. Receiving records, all lots, temperature records
b. Records review, calibration of thermo-recorders, histamine analysis of selected samples

6.3.1.5. HACCP Analysis of Heat-sterilized Fish Products (Canned fish)

The basis for canning is thermal processing of fish to achieve sterility in the final product. The canned fish are distributed at ambient temperatures and often stored for months and even years. The contents are normally eaten without any heating. Canned fish can cause of outbreaks of botulism and cases of histamine and staphylococcal enterotoxin poisoning.

Significant hazards identified are

a. Quality of raw material (biotoxins, chemicals)

b. Quality of cans

c. Corrective filling for proper heat-penetration

d. Faulty sealing result in recontamination

e. Survival of pathogens (*C. botulinum*) during heat processing

f. Recontamination of product after heat processing (faulty containers, poor sealing, contaminated cooling water, faulty container handling).

g. Contamination of hot and wet cans with *S. aureus*

Preventive measures are

a. Use of chlorinated cooling water is a safe precaution.

b. Isolation of the storage area of hot and wet cans and application of GHP by personnel

The CCPs identified are

a. Receiving step

b. Filling step

c. Sealing step

d. Retorting step

e. Cooling step

f. Post process handling

Critical limits are

a. Cans must meet container specifications for safety

b. Botulinum cook or 12-D process

c. Measure residual chlorine in the cooling water

Monitoring program

a. Letter of guarantee from supplier. Visual examination of all empty cans

b. Visual check of filling regularly (every half hour) by floor supervisor

c. Check the can closures at regular intervals (every half hour) visually. Do take down measurements at the beginning of the shift and every 2 hours by Q.C. staff

d. Put the products on hold for reprocessing, if time/temperature requirements violation, and identify the cause.

e. Test the samples at least two times per day by a designated person

Corrective actions

a. Reject defect cans. Contact supplier

b. If faculty sealing, shut down of processing line and inform plant manager. All products produced since last check must be put on hold. Identify the cause of the problem before starting up again.

Verification

a. Include a review of all operations - monitoring procedures and calibration of thermometers and automatic recorders

b. Additional Verification Procedures are common practice and, in some cases, a legal requirement. It includes checks to ensure that products have undergone appropriate heat treatment

   1. Incubation tests carried out at 37°C for seven days or at 35°C for ten days

   2. Microbiological examination of contents in the establishment or approved laboratory.

Records

a. Record any actions and measurements

6.3.1.6. HACCP Analysis of Cured Fishery Products - Lightly-Preserved Fish Products

Lightly preserved fish products are those with low salt content (<6%) and low acid content (pH >5.0) with or without the addition of preservatives such as sorbate, benzoate, NO_2, and smoke. The products are prepared either from raw or cooked raw material, but are normally consumed without any prior heating, e.g. salted, marinated, cold smoked or gravad fish. The products have a limited shelf life and are typically stored at temperature of 5°C.

Significant hazards identified are

a. Growth of pathogenic bacteria from the aquatic or the general environment

b. Production of biogenic amines (scombroid fish)

c. Presence of parasites

d. Chemical contamination (depending on geographical area).

Preventive measures are

a. Prevention of growth of C. botulinum by 3.5% salt and a storage temperature of 5°C
b. Reduce the shelf life of the products to a period of no growth of *Listeria monocytogenes.*
c. Prevent the growth of a number but not all of histamine producing bacteria by storage at low temperature (<5°C)
d. Introduction of a freezing step (-20°C for at least 24 h,)
e. Procure raw material from areas with no chemical contamination.

The CCPs identified are

a. Receiving step
b. Salting step
c. Freezing step

Critical limits

a. Receiving step: Use only raw material with good sensory quality. Do not allow fish from an area where there is a CFP advisory or area closed for fishing
b. Salting: WPS (water phase salt) >3.5% NaCl
c. Freezing step: -20°C for at least 24 hours
d. Storage temperature : <5°C

Monitoring programs

a. What: Sensory quality of raw material. Certificate of origin of fish. Salting procedures. Temperatures and times of freezing
b. How: Visual
c. When: All lots. Continuous recording of temperature
d. Who: Receiving employee. QC staff

Corrective actions

a. Reject lots of poor quality or with no certificate of origin
b. Adjust salting process
c. Check WPS in lots produced when process is out of control
d. Adjust freezing procedures.

6.3.1.7. HACCP Analysis of Fermented Fish

Fermented fish covers both enzymatically hydrolyzed and microbially fermented fish products. Fermented fish as "products which contain a carbohydrate source and in which the level of salt is less than 8% water phase salt (WPS)". This level of salt allows the fermentative growth of lactic acid bacteria and a concomitant decrease in pH to <4.5. In contrast, enzyme hydrolyzed fish has a WPS >8% and a final pH between 5-7. A large number of fermented fish products are found in South-East Asia. The products are traditionally stored at ambient temperatures and consumed without any cooking. Fermented fish are associated with a number of outbreaks of food borne diseases such as botulism, trematodiasis, salmonellosis and vibriosis.

Significant hazards identified are

a. Growth of C. botulinum type A and B, Listeria monocytogenes, Vibrio spp.
b. Pathogenic viruses
c. Biotoxins (ciguatera)
d. Biogenic amines (scombroid species)
e. Parasites such as trematodes

Preventive measures are

a. Rapid and adequate acidification. Reduce the temperatures during fermentation.
b. Control of viruses by prerequisite program
c. Food safety education. Bring about changes in the traditional consumption practices of eating non-cooked fermented fish.
d. Freezing for fermented fish eaten without any cooking

The CCPs identified are

a. Receiving step to check raw materials
b. Chilling step during fermentation to inhibit the growth of indigenous pathogens
c. Freezing step to control of parasites.

Critical limits

a. pH near pH 4.5.
b. Temperature <10°C for complete safety.

6.3.1.8. HACCP Analysis of Semi-preserved Fish

Semi-preserved products are those fish products with >6% water phase salt (WPS) or a pH <5.0 with or without added preservatives such as sorbate, benzoate, and nitrate. These products require chill storage (< 10°C) and have a shelf life of 6 months or more. No heat-treatment is applied neither during processing nor in the preparation before consumption. Traditional production includes a long ripening period (several months) of the raw material before final processing e.g. salted and marinated fish, fermented fish and caviar products. These products cause illness due to the presence of bacterial toxins, parasites, biotoxins and histamine.

Significant hazards are

a. Non-indigenous pathogenic bacteria (bacteria and viruses)

b. Bacterial toxins (botulinum)

c. Biotoxins (ciguatera)

d. Biogenic amines

e. Parasites

f. Chemical contamination

Preventive measures are

a. Storage temperature at <10°C prevent growth and toxin production by pathogenic bacteria

b. Full control over handling and processing steps from harvesting to consumption reduce botulinum

c. Avoid fish species with a CFP advisory or problem

The CCPs and critical limits are

a. Receiving step : Check raw material

b. Chilling step : Time-temperature conditions for prevention of growth of pathogens. Critical limits are < 5°C for raw materials and < 10°C for final products

c. Salting step : Critical limit is WPS 6% for killing parasites

d. Addition of acids and/or preservatives: Critical pH limit is 5.0

e. Freezing step: for killing of parasites.

Monitoring procedures, corrective actions, verification and record keeping are described as earlier.

6.3.1.9. HACCP analysis of Dried, Smoke-dried, Heavily-salted Fish

Dried or salted fish products have a very high salt content (>10% WPS) and/or a very low water activity (a_w 0.85). Dried or salted fish are stable at high temperatures and therefore stored and distributed at ambient temperatures. The most salt-tolerant pathogenic organism is Staphylococcus aureus that can grow at a_w 0.83 and produce toxin at a_w 0.85, and is a significant hazard. A critical phase in processing is the time taken for the penetration of salt until the WPS reaches 10% or a_w to below 0.85 in the thickest part of the fish. Larger fish are therefore eviscerated prior to processing.

Significant hazards are

a. Raw material quality (biotoxin, chemical contamination and histamine)

b. Contamination of dried or salted fish with enteropathogenic bacteria and viruses

Preventive measure is

a. Prerequisite program to control enteropathogenic bacteria and viruses.

The CCPs identified are

a. Receiving step

b. Salting/drying step

Critical limit

a. Critical limit is 10% WPS or a_w 0.85 in fish flesh.

6.3.1.9. HACCP Analysis of Mildly Heat-processed Fish Products

Pasteurized or cooked and breaded fish fillets, cooked shrimp and crabmeat, cook-chill products and hot smoked fish receive a heat treatment during processing, and then packed and stored/distributed, as chilled or frozen products. Some of them receive an additional heat treatment before consumption (cooked and breaded fillets, cook-chill products) or some are eaten without any further treatment (hot smoked fish, cooked shrimp). Some of them are ready-to-eat products extremely sensitive to contamination. Epidemiological evidence show causes of food poisoning is due to the growth of coagulase-positive Staphylococcus aureus and enteropathogenic organisms among Enterobacteriaceae and Vibrionaceae.

Significant hazards identified are

a. Survival of pathogens
b. Recontamination after cooking
c. Growth of pathogens
d. Raw material quality (chemical hazards).

Preventive measures

a. Avoid recontamination and growth of pathogens are taken care by the prerequisite program.
b. Check for faulty container seal or incorrect hot filling process

The CCPs identified are

a. Receiving step: Control of raw materials
b. Cooking step: Control of survival of pathogens

Critical limit

a. Cooking step to achieve 6D process to eliminate the target organism, Listeria monocytogenes

References

CAC 2003. Code of Practice for Fish and Fishery Products. Codex Alimentarius Commission of Food and Agriculture Organization / World Health Organization, Rome, Italy.

FDA 2001. Current Good Manufacturing Practices. 21 CFR Part 110. http://seafood.ucdavis.edu/GUIDELINES/gmps. htm, US Food and Drug Administration.

ICMSF 1988. Microorganisms in Foods 4. Application of the Hazard Analysis Critical Control Point (HACCP) system to ensure microbiological safety and quality. International Commission on Microbiological Specifications for Foods, Blackwell Scientific Publications.

NACMCF 1998. Hazard analysis and critical control point principles and application guidelines. Journal of Food Protection 61, 762-775, National Advisory Committee on Microbiological Criteria for Foods.

Ryder J., Karunasagar I. and Ababouch L. 2014. Assessment and management of seafood safety and quality: current practices and emerging issues. FAO Fisheries and Aquaculture Technical Paper No. 574. Rome, FAO. 432 pp.

7

Risk Assessment

Introduction

The risk assessment is gaining importance as the scientifically-based approach for the development of food safety and quality standards. It is a logical extension of HACCP revolution. In hazard analysis, first the hazards likely to occur in foods are identified, then an assessment is made of the severity of each hazard, and an evaluation is done of its likelihood to occur. Severity and likelihood are the two factors involved in risk assessment. Second, the increase in international trade has raised new safety and quality challenges. New approaches address the risk of cross-border transmission of infectious and hazardous agents with emerging food-borne diseases and quality problems. This has led to the development of a new safety and quality regulatory framework in 1995, the Sanitary and Phytosanitary (SPS) and the Technical Barriers to Trade (TBT) Agreements of the World Trade Organization (WTO). In these two Agreements, two provisions are of importance to fish safety and quality:

- National SPS and quality requirements should reflect standards agreed on in the international standards setting bodies i.e. Codex Alimentarius for food quality and safety.
- Domestic standards can be developed based on scientific risk assessment.

Risk assessment of microbiological hazards is identified as a priority area by the Codex Alimentarius Commission (CAC). In 1999, the Codex Committee on Food Hygiene (CCFH) in the 32nd session, identified a list of 21 pathogen-commodity combinations that require expert risk assessment advice. FAO and WHO jointly launched a programme with the objective to provide an expert advice on risk assessment of microbiological hazards in foods to their member countries and to the CAC. Under which, expert groups are established to examine four priority pathogens: product pairings, e.g. *Listeria monocytogenes* in ready-to-eat food; *Salmonella* in eggs and broiler chickens; *Campylobacter* spp. in broiler chickens; and *Vibrio* spp. in seafoods. Risk assessment is therefore important throughout all aspects of the seafood industry - for companies, national governments and for international regulators.

7.1. Basics of Risk Assessment

Risks from microbiological and chemical agents are of serious concern to human health. The definitions and terms used in risk analysis are set out in the CAC Principles and guidelines for the conduct of microbiological risk assessment (CAC/GL-30, 1999). Risk is a function of the probability of an adverse health effect and the severity of that effect is consequent of hazard (s) in food. Risk has two parts.

1. The likelihood of a hazard affecting us
2. The severity of its consequences

7.1.1. Risk Analysis

Risk analysis is a process consisting of three components

1. Risk assessment
2. Risk management
3. Risk communication.

The risk managers identify the need for a risk assessment and select an assessment team; and also begin the risk communication process.

7.1.2. Risk Assessment

Risk assessment is a scientific process consisting of the following steps:

1. Hazard identification
2. Hazard characterization
3. Exposure assessment
4. Risk characterization.

The aim of risk assessment is to estimate the level of illness expected in our target population from a product (s).

7.1.2.1. Hazard Identification

Hazard identification is the first stage of risk assessment. The biological, chemical and physical agents that cause adverse health effects in a particular food or group of foods are identified. For example, *Clostridium botulinum* is identified as a hazard in canned, smoked and vacuum-packed seafoods, but is unlikely to be a hazard for other products. So, hazard identification is a primary screening process that allows risk managers to eliminate product:pathogen pairs that are of no concern.

7.1.2.2. Hazard Characterization

Hazard characterization is the qualitative and/or quantitative evaluation of the nature of the adverse health effects associated with biological, chemical and physical agents present in food. In microbiological risk assessment, the concerns relate to micro-organisms and/or their toxins. There are two parts in hazard characterization.

1. A description of the effects of the hazard (micro-organism or toxin)
2. The dose-response relationship, if it exists.

Dose-response assessment

The dose-response assessment is a link between the amount of the hazard ingested (dose) and the chance of an individual getting infected and the scale of the illness. For example, most healthy individuals consume large numbers of *Listeria monocytogenes* (i.e. > 100 million cells) with no serious illness. But, in the susceptible group (pregnant women, the aged or immune compromised individuals), a much smaller dose (as few as 10000 cells) can cause serious illness and, even death, in 30% of cases.

7.1.3. Exposure Assessment

To perform an exposure assessment, data are required in two areas: (1) number of servings of potentially dangerous food; and (2) level of contamination with the micro-organism or toxin at the time of consumption. For which, the micro-organism or toxin is followed through the processing i.e. food preparation chain and the changes occur to the hazard is estimated.

7.1.4. Risk Characterization

Risk characterization is done by integrating hazard identification, exposure assessment and hazard characterization to provide a risk estimate. Risk estimate is an output of risk characterization. It can be either a qualitative estimate (high, low, medium) or quantitative estimate. The number of people expected to become ill from the particular product:hazard pairing can be predicted from the risk estimate. Risk characterization can also be semi-quantitative. Risk ranking is also given as a number in a specific range, for example, 0-100.

7.1.5. Risk Management

Risk management is the process distinct from risk assessment. Risk management process evaluates policy alternatives by considering risk assessment and other factors, to protect the consumer health and to promote fair-trade practices, and, if needed, to select appropriate prevention and control options. Risk managers must consider the views of interested groups viz. scientists, industry, consumer groups, politicians and lawyers.

7.1.6. Risk Communication

Risk communication is the interactive exchange of information and opinions throughout the risk analysis process concerning hazards and risks, risk-related factors and risk perceptions among risk assessors, risk managers, consumers, industry, the academic community and other interested groups. It also shares information on the explanation of risk assessment findings and the basis of risk management decisions. Risk communication is a very difficult task because it involves the full range of stakeholders. A major problem is to inform consumers that no food product is risk-free and, as a consequence, consumers must be prepared for X deaths and Y illnesses each year from that particular product.

7.1.7. Risk Profile

Risk profile is a description of a food safety problem. Risk profiling is made to identify the elements of a hazard or risk, that are relevant to risk management decisions. Risk profiling is the responsibility of the risk manager. Risk profiling is done for the seafood industry and there are differences in risk rating. For example, dried and heavily salted fish have no risk. But, during rainy season mold formation occurs and molds produce aflatoxin, and the risk rating is now not zero.

7.1.8. Uncertainty Analysis

Uncertainty analysis is a method used to estimate the uncertainty associated with model inputs, assumptions and structure/form. Risk assessments contain a statement specifying that insufficient data are available in one or more areas and, as a result, a certain amount of caution is attached in the risk estimate. Caution leads to uncertainty. One should always record the data gaps that lead to uncertainty. Later, when the knowledge becomes available, the level of uncertainty can be reduced, so that the risk estimate becomes more accurate.

7.2. Principles and Objectives of Risk Assessment

The CAC set out general principles and guidelines for the conduct of microbiological risk assessment in 1999 (FAO/WHO, 2001). The principles state that:

1. Risk assessment should be sound based upon science.
2. There should be functional separation between risk assessment and risk management.

3. Risk assessment should be conducted according to the four structured approach
4. Risk assessment should clearly state the purpose, including the form of risk estimate
5. Risk assessment should be conducted transparently.
6. Any constraints that impact on the risk assessment (cost, resources or time) should be identified and their consequences described.
7. Risk estimate should contain a description of uncertainty.
8. Data should determine uncertainty in the risk estimate; data and data collection systems should be of sufficient quality and precision to minimize uncertainty in the risk estimate
9. Microbiological risk assessment should consider the dynamics of microbiological growth, survival, and death in foods and the complexity of the interaction between human and agent following consumption, and the potential for further spread.
10. Risk estimates should be reassessed over time by comparison with independent human illness data, whenever possible
11. Risk assessment need re-evaluation, when new information is available.

7.3. Types of Risk Assessment

There are three types of risk assessments, namely

1. Qualitative risk assessment
2. Semi-quantitative risk assessment
3. Quantitative risk assessment

7.3.1. Qualitative Risk Assessment

Qualitative risk assessment is the simplest and quickest, but it is subjective. Every HACCP plan contains simple qualitative risk assessments in the HACCP worksheet (Table 7.1). For every hazard, a risk estimate is given as high, medium or low on the severity of the hazard and the likelihood of its occurrence. For example, the process step is retorting in fish canning and the hazard is *Clostridium botulinum*. Severity of the hazard is high, but likelihood is low, because billions of cans of fish are manufactured each year with no sign of the hazard.

Table 7.1. Hazard control worksheet

Process step	Hazard	What can go wrong	Risk		Hazard control
			Severity of hazard	Likelihood of hazard occurring	
	Biological				
	Chemical				
	Physical				

In qualitative risk assessment, the risk estimate is given as risk ranking - high, low and medium (Table 7.2).

Table 7.2. Qualitative risk ranking

Hazard	Product	Severity of hazard	Likelihood of occurrence	Exposure in diet	Linkage with epidemiology	Risk ranking

The risk assessment is based on factors linked with exposure assessment (likelihood) and with hazard characterization (severity). If the hazard: product pairing has some linkage with epidemiology (food poisonings), there is a probability of it happening again. For example, ciguatera occurs in two different populations. In a Pacific island atoll community, the likelihood of occurrence of ciguatera is high. In the population of the UK, the likelihood of occurrence is very low. Risk ranking is high or very high for the Pacific atoll and low or very low for the UK.

Another qualitative scheme for categorizing risk from seafoods is ascribing pluses to hazard, then rank risk as "high" (four or more pluses) or "low" (less than four pluses). The scheme considers epidemiology and the process, critical control point (CCP) for each hazard and assess the possibilities for growth and death of microbial hazards (Table 7.3).

Table 7.3. Qualitative risk assessment based on the process

Risk criteria		Raw molluscan shellfish	Canned fish	Dried fish
Bad safety		+	+	-
No CCP for		+	-	-
Possibility of contamination or recontamination++-		+	+	-
Abusive handling		+	-	-
Growth of pathogens		+	-	-
No terminal heating		+	+	+
	Risk category	High	Low	No risk

Molluscan shellfish, fish eaten raw, lightly-preserved fish and mildly heat-treated fish are considered "high" risk, while chilled/frozen fish and crustaceans, semi-preserved fish and heat-processed (canned) fish are considered "low" risk; and dried and heavily salted fish are considered to have no risk.

7.3.2. Semi-quantitative Risk Assessment

Semi-quantitative risk assessment is a numerical risk estimate based on a mixture of qualitative and quantitative data. A simple spreadsheet tool is used to describe the risk that emerges from pathogens in products produced by typical processes (canning, chilling, cooking, etc.). Table 7.4 lists the risk criteria needed for a semi-quantitative risk assessment. There are simple questions to be answered qualitatively in terms such as "high" and "low".

Table 7.4. Typical risk criteria in a semi-quantitative risk assessment

Risk criteria	Input
Dose and severity	
1. Hazard severity	
2. Susceptibility	
Probability of exposure	
3. Frequency of consumption	
4. Proportion consuming	
5. Size of population	
Probability of infective dose	
6. Probability of contamination	
7. Effect of process	
8. Possibility of recontamination	
9. Post-process control	
10. Increase to infective dose	
11. Effect of treatment before eating	

A tool named "Risk Ranger" is used to provide a risk profile for the Australian seafood industry. The most robust risk estimates from Risk Ranger are a risk ranking (score from 0 to 100) and the number of illnesses per annum.

7.3.3. Quantitative Risk Assessment

Qualitative risk assessment provides numerical expressions of risk and indication of the uncertainties. Quantitative risk assessments are done for specific purposes. The numerical risk estimates are to answer questions that are posed by the risk managers, who originally commissioned the assessment. There are many QRAs in the seafood area and the notable ones are

- *Listeria monocytogenes* in smoked fish in Sweden
- *Vibrio parahaemolyticus* in oysters in the United States
- *Listeria monocytogenes* in a range of seafoods in the United States

The risk assessments were taken in response to large outbreaks of food poisoning. In 1997 and 1998, there were two incidents involving *Vibro parahaemolyticus* in oysters that caused more than 700 cases of illness. In the late 1990s, there were two listeriosis incidents caused by *Listeria monocytogenes* in the United States involving hot dogs and delicatessen meats, in which more than 130 were seriously ill and 28 died. In the United States, risk assessments took a very large period of more than one year to prepare and then a 1-2 year review period of public comment. A typical QRA was carried out for smoked fish in Sweden indicated that the predicted annual number of illnesses varied between 47 and 2 800 for consumers at the most risk.

7.4. How to Preform Risk Assessments?

Risk assessment is carried out with people, information and data handling. There are different tasks to perform the risk assessments

7.4.1. Process Initiation

The task in process initiation i.e. responding to customers' requirements, is done at four stages, which are given below:

Stage 1 - Assemble a team

A risk assessment team is constituted as that of HACCP team and it consists of

1. A seafood technologist with knowledge of processes and products
2. A food microbiologist, who knows about microbial ecology
3. A statistician to assemble and handle data
4. A manager to direct the work

The risk manager should find specialists to undertake the risk assessment work.

Stage 2 - Survey the industry

The risk assessment team makes a survey of all seafood products manufactured within the country and the importing countries. This is a straightforward task as every nation keeps a record of its seafood production volumes and species.

Stage 3 - Survey seafood related illness

The risk assessment team conduct a preliminary study of seafood related illnesses within the country and in the importing countries. The survey can begin from the Health Department, who keeps the record of food poisoning incidents. A list of seafood incidents is made by linking products with hazard (microorganisms, toxins, or chemicals) and include these hazards: product pairs in the risk profile. The next stage is collection of statistics from the importing countries. There are a number of websites that give information on food poisonings (Table 7.5).

Table 7.5. Sources of information on seafood illness and recalls of seafoods

Country	Organization	Website
European Union	Eurosurveillance Weekly	http://www.eurosurv.org
USA	Center for Science in the Public Interest	http://www.cspinet.org
USA	Morbidity and Mortality Weekly	http://www.cdc.gov/mmwr
UK	Public Health Laboratory Service	http://www.phls.nhs.uk
Australia	Communicable Diseases Intelligence	http://www.health.gov.au
Australia	Food Standards Australia and NZ	http://www.anzfa.gov.au
International	Food Safety Network	http://www.foodsafetynetwork.ca

Once all data are gathered, assemble them into a summary table. The data are helpful to identify the main seafood hazards. The data provide background information on the problems caused in importing countries. The list of hazards and products form the base of risk profiling (Table 7.6). Examples are mercury in swordfish and sulfite and chloramphenicol in shrimp.

Table 7.6. Hazards and products in the risk profile

Hazards	Products
Chemical Hazards	
Ciguatera	Reef fish
Mercury	Predaceous fish
Sulfite	Shrimps
Biotoxins	Bivalve molluscs
Biological Hazards	
Viruses	Bivalve molluscs
Listeria monocytogenes ?	Smoked seafood
Vibrio parahaemolyticus	Shellfish eaten raw
Staphylococcus aureus	Cooked seafood
Clostridium botulinum	Canned, vacuum packed seafood
Histamine	Scombroid fish
Parasites	Raw fish

Stage 4 – Do a risk profile

Risk profiling is a description of a food safety problem and the framework developed to identify the elements of a hazard or risk relevant to risk management decisions. The data are gathered in three areas, to identify pathogen:product pairing for further investigation.

1. Hazard identification
2. Hazard characterization
3. Exposure assessment

7.4.1.1. Hazard Identification

For hazard identification, the following are identified for each hazard:product pairing.

1. Links with confirmed food borne illnesses within the country and in importing countries
2. International food borne disease outbreaks
3. Recalls monitored by food authorities in importing countries

The above information gives an idea of the food safety relevance of the hazard:product pairing on whether to have it in the risk profile. If any selected hazard:product pairing has not caused any problem, it need not be included in a risk profile. For example, parasitic worms do not cause problem in finfish exported as frozen fillets, as freezing kills the parasites. There is a CCP (freezing step) to eliminate the hazard, and the risk.

7.4.1.2. Hazard Characterization

Hazard characterization is composed of inter-relationships between the food matrix, the host, and the hazard (pathogen, toxin, or chemical agent), with that of the disease. This is a very complex part of risk assessment. In a large-scale food poisoning, only a proportion of consumers become ill and of which, only a small proportion will die. The reason is more complex, but we have to identify the reasons of a large outbreak. There are several main factors to be considered to do the hazard characterization.

1. The Host : For example, the vulnerable individuals for the outbreak of listeriosis from Mexican cheese are identified as pregnant women, fetuses, newly born babies, immunocompromised individuals, and elderly.
2. The Pathogen: *Listeria monocytogenes* has several properties that allow it to infect particular groups, as it can cross the placenta and infect the foetus; and the number of cells needed to cause listeriosis is much lower for vulnerable groups.

3. The Food Matrix: Mexican cheese give certain competitive advantages for the microbial ecology of *L. monocytogenes, L. monocytogenes* is tolerant to the salt levels in cheese while competing bacteria are inhibited by salt. *L. monocytogenes* has the ability to grow in the refrigerator. *L. monocytogenes* is protected within curd particles or fat, while passing through the stomach of the consumer.

Consumers (Hosts)

It is important to assess why some people become infected more readily than others? and how many consumers fall into at-risk groups? All individuals lose immunity either progressively with age or catastrophically, if they undergo chemo-or radiotherapy or if they acquire disease that impair the immune system. To calculate the proportion of vulnerable population to infection, national statistics can be accessed. If national statistics do not exist, use 20% as a reasonable indication.

a. **Aging :** National statistics give the number of elderly and aged people in the community. Antibody levels are high in childhood. At the age of 55, the levels reduce to 50% and at the age of 90, only 25% remains. The elderly has a reduced neutrophil function, which reduce intracellular antimicrobial activity. The loss of immunity and antimicrobial activity in the elderly is a concern.

b. **Acquired immune deficiency syndrome:** Individuals infected with human immune-deficiency virus (HIV) are at increased risk of gastro-enteric infections and Salmonella infection, in particular.

c. **Cancer:** Lung, breast, bowel, lymphoma, leukemia and renal cancers can increase an individual's susceptibility to food borne diseases. It is very difficult to estimate those undergoing therapy. If the median period of treatment is taken as five years, there is a quite a large number within the population.

d. **Diabetes:** In developed countries, 10% of the population are insulin dependent. Non-insulin dependent diabetes is more common in older individual, who are overweight and sedentary. A similar number of undiagnosed asymptomatic diabetics also exists. Diabetes is common among indigenous and Pacific island groups at the rate of 10%.

e. **Pregnancy:** There are two phases of pregnancy during which mother or foetus are at greater risk. In the first trimester, fetuses are at risk from the effect of heavy metals. In the third trimester, both of them are susceptible to *L. monocytogenes*.

f. **Very young:** Children less than five years old have a greater risk of food-borne illnesses. The prevalence of Salmonellosis among children less than 6 months old is due to low gastric acidity, immature immune response and low protective effect from residual gut microflora.

g. **Hypochlorhydria:** In many western countries, people use preparations to reduce stomach acidity, by taking acid-lowering agents, which range between 1-5%.

7.4.1.2.1. Dose-response Measurement

Dose-response is a measure of the quantity of disease agent required to cause disease. For example, how many *Salmonella* cells in a meal of cooked shrimp are need to give Salmonellosis? How many Hepatitis A virus particles in a meal of raw oysters needed to contract hepatitis? How much ciguatoxin in reef fish is required to get ciguatera poisoning? The ability of the specific microorganism or toxin or chemical agent becomes important – its virulence or pathogenicity or toxicity. A range of studies can be conducted to collect information on these aspects.

1. **Volunteer feeding studies:** A straightforward way of finding the quantity required to cause illness. In early days, men serving prison sentences are fed with meals containing different levels of *Salmonella*, and at least 10^5 cells are needed to cause illness. The changes in the society has now reduced volunteer feeding studies.

2. **Epidemiological data:** The leftover food is tested after an outbreak of food poisoning to find out the causative organism and its population. The test results give fairly the number of organisms needed to produce illness. For example, a person needs to eat $>10^5$ cells of *Salmonella* to become ill, but it is not always true, because in a number of outbreaks, only a few has caused infections. The food matrix, especially its fat content is important. Foods with high fat contents protect the *Salmonella* from gastric juices.

   In some other outbreaks, the numbers needed for illness are much higher than previously thought. For example, the level of *L. monocytogenes* in mussels was $>10^7$ cells/g, which suggested more than 100 million cells are consumed in each meal, which had caused illness due to gastro-enteritis but not listeriosis, in people who ate smoked mussels in Australia.

3. **Animal studies:** Laboratory animals are used to determine the disease-causing agent needed to cause symptoms. Mice is widely used, as they are cheap to raise and feed. Primates and pigs are expensive. Main disadvantages are that the response is different from that of humans and in many countries, there is ethical opposition for the use of animals. Mouse

injection is important in studying the toxin of *Clostridium botulinum*, one of the dangerous microorganisms associated with canned seafood.

4. ***In-vitro* studies:** Cell lines are maintained in cell culture to test toxins and microorganisms under controlled conditions. The major limitation is that it is difficult to relate the findings to human dose-response.

The data collected on dose-response of different disease-causing agents associated with seafood are summarized (Table 7.7). The levels required to cause illness are an indication of the relative toxicity of seafood agents (from very high toxicity of botulinum toxin to relatively low toxicity of histamine). There is a great disparity between levels required to infect susceptible versus non-susceptible individuals.

Table 7.7. Ranges of agents associated with seafood needed to cause illnesses

Agent	**Susceptible groups**	**Non-susceptible groups**
Toxins		
Ciguatera	Approx. 50/ng	Approx. 1ng/kg bodyweight
Histamine	Approx. 50 /mg	Approx. 1 mg/kg bodyweight
Paralytic shellfish poison	150-1500ug	150-1500 ug
C. botulinum toxin	0.1-1.0 ug	0.1-1.0 ug
Microorganisms		
Salmonella	10-100 cells	100000 cells
Vibrio parahaemolyticus	10000 cells	10000 cells
Listeria monocytogenes	1000 – 10000 cells	1000000 cells
Hepatitis A virus	1-10 particles	10-100 particles

The information on hazard characterization for each of the disease-causing agents are available i.e. virulence and infectivity of various consumer groups (vulnerable and non-vulnerable), illness caused (time of onset, duration, symptoms), sequelae (ability to cause further conditions), effect of food matrix (composition, processing, meal preparation) on the agent.

7.4.1.2.2. Dose-response Models

Risk modelers are integral part of risk assessment. Risk modelers think differently than microbiologists and food technologists. Modelers are important to do quantitative risk assessment. There are several models surrounding dose-response in FAO/WHO guidelines on hazard characterization for pathogens in food and water.

7.4.1.3. Exposure Assessment

The exposure to a disease-causing agent depends on three factors:

1. The level of the agent in the meal
2. The amount we eat (i.e. serving size)
3. The frequency with which people consume that component

For example, ciguatera in reef fish. People live on a Pacific atoll eat seafood every day, including species such as Spanish mackerel, a reef fish and consume up to 250 g of finfish in one serving and in total, consume 50 kg of reef fish every year. People from European city, consume seafood once a week and the serving size is around 100 g once a year. The exposure to ciguatera is very different i.e. 500 times. An exposure assessment is more complicated because a large number of other factors have to be considered such as :

1. Frequency of contamination (prevalence) with toxin or pathogen
2. Changes in level of contamination through the marketing chain
3. Seasonal effects
4. Consumption patterns
5. Susceptibility of consumer
6. Preparation effects

The better the exposure assessment, the more valid will be the risk estimate. The availability of local data is very important for exposure assessment.

7.4.1.4. Risk Characterization

In risk characterization, information from hazard identification, hazard characterization and exposure assessment are integrated to given the risk estimate. It is an estimate that shows the number of people who become ill, and the severity of illness, in respect to the specific agent in the product. If a qualitative risk assessment, the risk estimate is a simple statement that the risk is high or low or medium. If a quantitative risk assessment, the risk estimate is a number, such as predicted illnesses per annum in the population or the probability of becoming ill from eating a serving of the product.

The main reason for doing risk assessment is that risk manager can use the output – the risk estimates in the characterization. The managers need to know about the uncertainty and variability in the risk estimate. For example, in smoked and gravid trout in Sweden, the estimate showed the number of annual cases by two dose response models. Method 1 predicted a mean of 168 cases and a range of 47-2800 cases. Method 2 predicted a mean of 95000 cases with a range of 34000 – 1600000. The difference in the ranges reflect the uncertainty built into the predictions.

Risk characterization is invaluable for risk managers in a sensitivity analysis. The analysis ranks the influence of each parameter on the risk. Some factors increase the likelihood of illness, while other decrease it. The probability to eating smoked or graved fish are most affected by

1. Level of contamination (number of *L. monocytogenes*)
2. Prevalence of contamination (percentage of serving contaminated)
3. Serving size (the more you eat, the more you are likely to become ill)
4. Proportion of virulent strain of *L. monocytogenes*

This helps the risk manager to focus on area that should receive priority action. If the assessors identify uncertainty within these areas, the manager decides to invest in studies to obtain better data and reduce the uncertainty. Some risk assessments present the sensitivity analysis as a chart with bars representing the extent of the impact of each parameter has on risk. Fig. 7.1 shows a typical chart. It is called a "Tornado chart" because of its shape. Each bar reflects to a particular property, which is correlated either with increased or decreased risk.

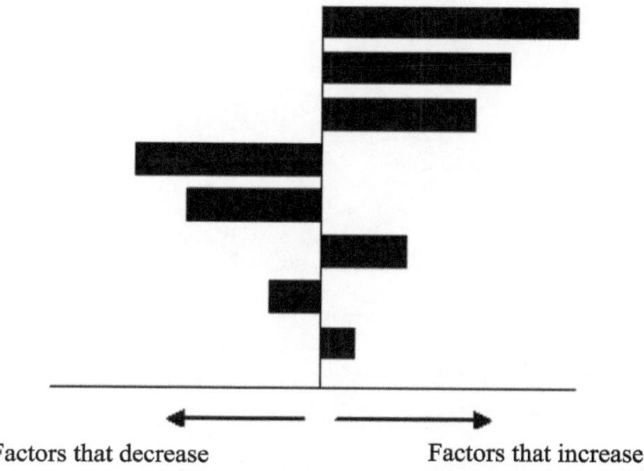

Fig. 7.1. Tornado chart showing the effect of each parameter on risk

After the risk assessment, a reality check has to be done to compare the predictions of annual illness with statistics of the Health Department of any country.

7.5. Examples of Risk Assessment Done

1. Qualitative risk assessment – Mercury in Fish
2. Semi-quantitative risk assessment – Ciguatera in reef fish, Histamine fish poisoning
3. Quantitative risk assessment – *Vibrio parahaemolyticus* in oysters

References

FDA, 2020. Fish and Fishery Products Hazards and Controls Guidance. Fourth Edition, U.S. Department of Health and Human Services Food and Drug Administration Center for Food Safety and Applied Nutrition, USA. www.FDA.gov/Seafood

Huss H.H., Ababouch L. and Gram L. 2003. Assessment and Management of Seafood Safety and Quality. FAO Fisheries Technical Paper 444. Food and Agricultural Organization of the United Nations, Rome. ISBN 92-5-104954-8

Sumner, J., Ross, T and Ababouch, L. 2004. Application of risk assessment in the fish industry. FAO Fisheries Technical Paper. No. 442. Rome, FAO. 78p.

8

International and National Food Laws and Standards, Regulatory Framework for Fish Safety and Quality

Introduction

In all countries, food is governed by a complexity of laws and regulations, which set out the Government requirements to be met by food chain operators to ensure food safety and quality. Food laws and regulations also govern standards that ensure the safety, quality, composition and suitability of food for consumers. Food laws and regulations of any nation depends on whether a nation adopts International norms developed by the Codex Alimentarius Commission (CAC) of the Food and Agriculture Organization (FAO) of the United Nations as well as the World Health Organization (WHO).

8.1. Food Laws

Food law means legislation that regulates the production, trade and handling of food. It covers the regulation of food control, food safety, quality and relevant aspects of food trade across the entire food chain, from the production to the consumer. Food laws protect the consumers against unsafe, adulterated and misbranded food and facilitate the movement of food across borders to sustain a fair trade.

Each nation administers its own laws, regulations and standards. It is also not static always and is constantly revised and updated in all nations. The food business operators (FBO) must obtain up-to-date information on any changes made in legislation, that are likely to affect the product standards. The regulations of all nations shall be obtained from their respective websites of the national ministries.

8.1.1. International Legislations

8.1.1.1. United States of America (USA)

In USA, the Food and Drug Act (1906) and the Federal Meat Inspection Act (1906) are the first federal food laws passed to protect American consumers. Later, Food, Drug and Cosmetic Act (1938) regulated the meat and poultry products. At present, Food Safety Modernization Act (FSMA), 2011 regulates all food products. https://www.fda.gov/food/guidance- regulation- food- and- dietary- supplements/ food- safety- modernization- act- fsma. Manufacturers play important role in preventing food borne illness. The Act gave the Food and Drug Administration (FDA) the authority to oversee and enforce supply chains. The Act shifts the focus towards prevention of food borne illness and insists every FBO to register with the FDA. The FDA finalized seven major rules to implement the FSMA. They are :

1. **Preventive Controls for Human Food** – Human food facilities registered with the FDA must implement a written plan that identifies hazards and outlines appropriate preventive controls
2. **Preventive Controls for Animal Food** – Animal food facilities registered with the FDA must implement a written plan that identifies hazards and outlines appropriate preventive controls
3. **Produce Safety** – Establishes minimum standards for growing, harvesting, packing, and storing produce
4. **Foreign Supplier Verification Program** – Importers must verify that their global suppliers comply with FDA regulations
5. **Third-Party Certification** – Accredits third-party certification bodies to administer voluntary consultative and regulatory audits to help companies prepare for regulatory audits or achieve certifications
6. **Food Defense (intentional adulteration)** – Food facilities registered with the FDA must develop a plan that assesses contamination vulnerabilities and document a mitigation strategy for each vulnerability
7. **Sanitary Transportation** – New requirements for companies that transport food, including shippers, receivers, loaders, and carriers.

8.1.1.2. European Union (EU)

The General Food Law (GFL) passed in 2002 provides a single framework for European Union. The aim of GFL is to protect public health and consumers in relation to food. It lays down the general principles governing food and feed in general. The GFL applies to all stages of food production, processing and distribution. The FBO is responsible to ensure that foods satisfy the requirements of food law, Regulation (EC) No 178/2002 of the European Parliament. https://eur-lex.europa.eu/legal-content/EN/TXT/?uri=OJ:L:2002:031:TOC. European Food Safety Authority (EFSA) is the keystone of European Union (EU) risk assessment regarding food and feed safety. http://www.efsa.europa.eu/en/aboutefsa.htm. The legal acts of EU are managed through regulations, directives, decisions, recommendations and opinions. Some of the important EU legislations related to food safety issues of fish and fishery products are as follows:

- Regulation (EC) No 178/2002: General principles and requirements of food law, establishing the European Food Safety Authority and laying down procedures in matters of food safety Regulation (EC) No 852/2004: Hygiene of foodstuffs.
- Regulation (EC) No 853/2004: Specific hygiene rules for food of animal origin
- Regulation (EC) No 854/2004: Specific rules for the Organization of official controls on products of animal origin intended for human consumption
- Regulation (EC) No 2073/2005: Microbiological criteria for foodstuffs
- Regulation (EC) No 882/2004: Official controls performed to ensure the verification of compliance with feed and food law, animal health and animal welfare rules
- Regulation (EC) No 1881/2006: Maximum levels for certain contaminants in foodstuffs
- Regulation (EC) No 333/2007: Methods of sampling and analysis for the official controls for the levels of lead, cadmium, mercury, inorganic tin, 3- MCPD and benzo(a)pyrene in foodstuffs
- Regulation (EC) No 1883/2006: Methods of sampling and analysis for the official control of levels of dioxins and dioxin-like PCBs in certain foodstuffs
- Regulation (EC) No 396/2005: Maximum residue levels of pesticides in or on food and feed of plant and animal origin
- Council Directive (96/23/EC): Measures to monitor certain substances and residues thereof in live animals and animal products
- Commission Decision (2005/34/EC): Harmonized standards for the testing for certain residues in products of animal origin imported from third countries
- Commission Decision (2002/657/EC): Implementing Council Directive 96/23/EC concerning the performance of analytical methods and the interpretation of results

- Commission Decision (98/179/EC): Official sampling for the monitoring of certain substances and residues thereof in live animals and animal products
- Commission Decision (2004/432/EC): Approval of residue monitoring plans submitted by third countries in accordance with Council Directive 96/23/EC
- Council Directive (96/22/EC): Prohibition on the use in stock farming of certain substances having a hormonal or thyrostatic action and of beta agonists
- Regulation (EC) No 470/2009: Community procedures for the establishment of residue limits of pharmacologically active substances in foodstuffs of animal origin
- Commission Regulation (EU) No 37/2010: Pharmacologically active substances and their classification regarding maximum residue limits in foodstuffs of animal origin
- Commission Regulation (EC) No 2023/2006: Good manufacturing practice for materials and articles intended to come into contact with food
- Commission Regulation (EC) No 1935/2004: Materials and articles intended to come into contact with food
- Commission Regulation (EU) No 1129/2011: Amendment to Annex II to Regulation (EC) No 1333/2008 of the European Parliament and of the Council by establishing a Union list of food additives
- Commission Regulation (EC) No 1333/2008: Food Additives
- Commission Regulation (EC) No 1334/2008: Flavorings and certain food ingredients with flavoring properties for use in and on foods
- Commission Regulation (EC) No 1331/2008: Establishing a common authorization procedure for food additives, food enzymes and food flavorings
- Directive 2000/13/EC: Labelling, presentation and advertising of foodstuffs (until 12 December 2014)
- Commission Regulation (EU) No 1169/2011: Provision of food information to consumers, amending regulations
- Commission Regulation (EU) No 1379/2013: Common organization of the markets in fishery and aquaculture products

8.1.1.3. Canada

The Canadian Food Inspection Agency (CFIA) develops and delivers programs and services designed to protect Canadians from preventable food safety hazards, to ensure that food safety emergencies are effectively managed, and that the public is aware of and contributes to food safety. http://www.inspection.gc.ca/english/toce.shtml.

8.1.1.4. Norway

Norwegian Food Safety Authority (NFSA) is a governmental body. Its goal is that consumers should have healthy and safe food and safe drinking water. It promotes human, plant, fish and animal health, environmentally friendly production, and ethically acceptable farming of animals and fish. The NFSA also performs duties relating to cosmetics and medicines, and inspects animal health personnel. https://www.mattilsynet.no/language/english/

8.1.1.5. United Kingdom

Food Standards Agency (FSA) is an independent food safety system set up by an Act of Parliament in 2000 to protect the public's health and consumer interests in relation to food. http://www.foodstandards.gov.uk/

8.1.1.6. Australia and New Zealand

Food Standards Australia and New Zealand (FSANZ) is a statutory authority in the Australian Government Health Department. It develops food standards for Australia and New Zealand. The Code is enforced by state and territory departments, agencies and local councils in Australia. The Ministry for Primary Industries in New Zealand and the Australian Department of Agriculture and Water Resources for food imported into Australia. https://www.foodstandards.gov.au/ .

8.1.1.7. Japan

There are four major laws in Japan pertaining to food safety and standards - the Food Safety Basic Law (2003), Food Sanitation Law, Japan Agricultural Standards Law, and Health Promotion law. The Food Safety Basic Law set the principles for developing a food safety regime and also set up the role of the Food Safety Commission, a food related risk assessment body. The Food Sanitation Law ensures the safety and sanitation of foods through the Ministry of Health, Labor and Welfare (MHLW), a food risk management agency. The law prohibits the sale of foods containing harmful substances. It also prescribes the standards for foods, additives, food containers and packages. The law is available in English on the following Japan External Trade Organization (JETRO) website: http://www.jetro.go.jp/en/market/regulations/.

8.1.2. National Legislations

In India, the Food Safety and Standards Act (2006) is passed by Food Safety and Standards Authority of India (FSSAI) under the Ministry of Health &

Family Welfare, Government of India. It lays down standards for articles of food and regulates manufacturing, processing, distribution, sale and import of food to ensure safe and wholesome food for human consumption. The Food Safety and Standards Regulations (FSSR) came into force in 2011. The FSS Act is formed by amalgamating seven old Acts and Orders on food safety handled by various Ministries and Departments such as Prevention of Food Adulteration Act, 1954; Fruit Products Order, 1955; Meat Food Products Order, 1973; Vegetable Oil Products (Control) Order, 1947; Edible Oils Packaging (Regulation) Order 1988; Solvent Extracted Oil, De- Oiled Meal and Edible Flour (Control) Order, 1967; and Milk and Milk Products Order, 1992. Website: http://fssai.gov.in/home.

The FSSAI has its headquarters at New Delhi and 6 regional offices located in Delhi, Guwahati, Mumbai, Kolkata, Cochin, and Chennai. There are 14 referral laboratories notified by the FSSAI, 72 State/UT laboratories and 112 laboratories are NABL accredited private laboratories notified by FSSAI. There are 21 regulations formulated for food safety and standards, and 5 for food safety and standards authority. Some of the important regulations are:

1. FSS (Licensing and Registration of Food businesses) Regulation, 2011
2. FSS (Packaging and Labeling) Regulation, 2011
3. FSS (Food product Standards and Food Additives) Regulation, 2011 (part I)
4. FSS (Food product Standards and Food Additives) regulation, 2011 (part II)
5. FSS (Prohibition and Restriction on sales) Regulation, 2011
6. FSS (Contaminants, Toxins and Residues) Regulation, 2011
7. FSS (Laboratory and Sampling Analysis) Regulation, 2011

8.2. International Regulatory Framework for Fish Quality and Safety

The growing demand for fish and fishery products and the increasing trend in international fish trade have raised major concerns about the overexploitation of aquatic resources and the quality and safety of the products. The following international regulatory framework for fish safety and quality assurance are available in order to harmonize the different fish safety and quality assurance approaches existing worldwide.

8.2.1. The World Trade Organization (WTO) Agreements

The WTO is the only global international organization dealing with the rules of trade between nations. The WTO agreements negotiated and signed by majority of the world's trading nations and ratified in their parliaments are the major achievements. The goal is to help producers of goods and services, exporters, and importers conduct their business. The WTO came into force in 1995. The WTO is the successor to the General Agreement on Tariffs and Trade (GATT) established in the wake of the Second World War. Article 20 of the GATT allows governments to act on trade in order to protect human, animal or plant life or health, provided they do not discriminate or use this as disguised protectionism. In addition, there are two specific WTO agreements dealing with food safety and animal and plant health and safety, and with product standards (1) Sanitary and Phytosanitary (SPS) agreement; and (2) Technical Barriers to Trade (TBT) Agreement.

8.2.1.1. SPS Agreement

The SPS agreement is meant to protect human or animal health from diseases caused by animals, plants, or products thereof; and from the risks of unsafe additives, contaminants, toxins, or disease-causing organisms in foods, beverages, or feedstuffs. The purpose of SPS Agreement is to ensure that measures are consistent without any arbitrary or unjustifiable discrimination on trade between countries. The agreement encourages the members of WTO to consider international standards, guidelines, and recommendations adopted by the Codex Alimentarius Commission (CAC) while establishing their own domestic SPS measures, for harmonization. This is guided by the WTO Committee on SPS measures having representatives of the CAC, the International Office of Epizootics (OIE), and the International Plant Protection Convention (IPPC). The OIE is now known as the World Organization for Animal Health (WOAH). The SPS measures are based on risk assessment carried out using internationally accepted risk assessment techniques. Risk assessment must consider the available scientific evidence, the relevant processes and production methods, the inspection/sampling/testing methods, the prevalence of specific illnesses and other matters of relevance.

8.2.1.2. TBT Agreement

The TBT agreement is a revision of the agreement first developed under the Tokyo round of negotiations (1973 - 1979). The objective of TBT agreement is to prevent the use of national or regional technical requirements, or standards in general, as unjustified technical barriers to trade. The TBT agreement emphasis on the use of standards, technical regulations, and assessment procedures set by international bodies, on terminology, packaging, marking, labeling, methods of analysis, product content regulation requirements for products, processes or production methods, to offer protection to the consumer against deception and economic fraud. Unlike the SPS Agreement, the TBT Agreement does not name international standard setting bodies, whose standards are to be used as benchmarks for judging compliance with the provisions of the Agreement.

8.2.2. USFDA Standards

The USFDA operates a mandatory safety program for all fish and fishery products under the provisions of the Federal Food, Drug and Cosmetic (FD&C) Act, the Public Health Service Act, and related regulations. The FDA program includes research, inspection, compliance, enforcement, outreach, and the development of regulations and guidance. The FDA publishes the Fish and Fisheries Products Hazards and Controls Guidance as follows in Table 8.1.

Table 8.1. The USFDA Fish and Fisheries Products Hazards and Controls Guidance:

Product	**Level**
Ready to eat fishery products (Minimal cooking by consumer)	*Listeria monocytogenes*- presence of organism in 25g sample
All fish	*Salmonella* spp. - presence of organism in 25g sample
All fish	*Staphylococcus aureus* - positive for Staphyloccocal enterotoxin (or) *Staphylococcus aureus*-level equal to or greater than 10^4 /g (MPN)
Ready to eat fishery products (Minimal cooking by consumer)	*Vibrio cholera*- presence of toxigenic O1 or non-O1 and non- O139 in 25g sample
Ready to eat fishery products (Minimal cooking by consumer)	*Vibrio parahaemolyticus* - levels equal to or greater than 1 x 10^4/g (Kanagawa positive or negative).
Post-Harvest Processed Clams, Mussels, Oysters, and Whole and Roe-on Scallops, Fresh or Frozen, that make a label claim of "Processed to reduce *Vibrio Parahaemolyticus* to Non-Detectable Levels"	*Vibrio parahaemolyticus* - levels less than 30/g (MPN).
Cooked ready-to-eat fishery products (Minimal cooking by consumer)	*Vibrio vulnificus* - presence of organism
Post-harvest processed clams, mussels, oysters, and whole and roe-on scallops, fresh or frozen, that make a label claim of "processed to reduce *Vibrio vulnificus* to non-detectable levels"	*Vibrio vulnificus* - levels less than 30/g (MPN).
All fish	*Clostridium botulinum* 1. Presence of viable spores or vegetative cells in products that will support their growth; or 2. Presence of toxin.

Product	Level
Clams, oysters, mussels, and whole and roe-on scallops, fresh or frozen	Microbiological 1. *E. coli* or Fecal Coliform - 1 or more of 5 subs exceeding MPN of 330/100 g or 2 or more exceeding 230/100 g 2. APC - 1 or more of 5 subs exceeding 1,500,000/g or 2 or more exceeding 500,000/g.

8.3. Food Standards

A standard is a document that provides requirements, specifications, guidelines or characteristics used to ensure that food is suitable for human consumption. A set of criteria that a food must meet including source, composition, appearance, freshness, permissible additive, microbial load and chemical contaminants. Food safety standards are broadly classified as mandatory and voluntary standards.

1. **Mandatory Standards:** Mandatory standards are the standards incorporated into a legislation and require compliance by law, regulation, government statute policy, or contractual agreement. In case of failure to comply with a mandatory standard, it results in civil or criminal penalties. Mandatory standards specify minimum requirements for the products to meet before supplying to the customers in order to prevent or reduce the risk. e.g. EU standards, USFDA standards

2. **Voluntary standards:** Voluntary standards are standards published in consensus with the industry. They contain specifications and procedures designed to ensure products, services and systems are safe, reliable and consistent. They address a range of issues even that are not food safety related. There is no legal action if the products do not meet voluntary standards e.g. Codex Standards and ISO Standards

8.3.1. International Standards for Seafood

Food business operators must comply with the standards imposed by the importing nation, where the food is consumed. The company intending to export its seafood must comply with both the domestic and foreign laws. To ease this process, nations began to accept international food standards. International standards are likely to impact domestic standards. If a nation participating in the world food trade, the companies must comply with the emerging voluntary international standards. The goal for the multi-lateral agreements is to achieve identical standards, i.e. at least harmonized standards. International standards are emerging from the Food and Agriculture Organization (FAO), the World Health Organization (WHO) of the United Nations and the World Trade Organization (WTO).

8.3.1.1. The Codex Alimentarius Commission (CAC)

The CAC was created in 1963 by FAO and WHO to develop food standards, guidelines and related texts such as Codes of Practice under the Joint FAO/WHO Food Standards Programme. The WHO of the United Nations is a specialized agency health established for on 7 April 1948 with the objective provide the highest possible level of health to all people. The FAO of the United Nations established in16 October 1945 takes international efforts to fight hunger and serve as a source of knowledge and information.

The main purposes are to protect health of the consumers and to ensure fair trade practices in the food trade, and to promote coordination of all food standards work undertaken by international governmental and non-governmental organizations. The CAC is an inter-governmental body with a membership of 189 comprising of 188-member countries and 1-member organization (the European Union). In addition, observers from international scientific organizations, food industry, food trade and consumer associations also attend sessions of the Commission.

The purpose of Codex is to develop international food standards, which are agreed and adopted on a world-wide, regional or group-of-countries basis. The Codex promotes acceptance and implementation of its standards by national governments. In countries where national or domestic standards are available, adoption of Codex standards ensures international acceptance.

The scope of Codex is to produce standards for all foods, which are processed, semi-processed or raw. The Codex documents have provisions in respect of food hygiene, food additives, pesticide residues, contaminants, labelling and presentation, and methods of analysis and sampling. The Codex provides numerical standards, codes of practice, and other guidelines through its committees. The Codex has committees with specialists to deal with separate areas. There are two major committees viz. commodity and general subject committees.

- **General Subject Committee**: Nine committees that deal with general principles, hygiene, veterinary drugs, pesticides, food additives, labelling, methods of analysis, nutrition and import/export inspection and certification systems.
- **Commodity Committee**: Twelve committees that deal with a specific type of food class or group, such as dairy and dairy products, fats and oils, or fish and fish products.

The work of the Committees on hygiene, fish and fishery products, veterinary drugs and import/export inspection and certification systems are important to ensure the safety and quality of internationally traded fish and fishery products.

The Commodity committee for Fish and Fish Products is functioning from Norway. The General Subject Committees function at USA (hygiene), Canada (labelling), the Netherlands (additives & pesticides), Hungary (analysis and sampling) and France (general principles). The member government takes the responsibility of each committee for piloting standards. The Codex standards are meant to be voluntary and adopted by consensus. Under the new SPS/TBT agreements, the Codex standards cannot be called "voluntary", nor fully mandatory.

8.3.1.1.1. Codex Standards for Fish and Fishery Products

The comprehensive minimum standards are available for a wide range of fishery products, meant for direct sale to the consumer. They are drafted by many countries collaborating with each other. The standard must pass through 10-step procedure to form a 'Recommended International Standard'. In step 9, the final standard is submitted to governments for their formal acceptance. The 'recommended standard' is then published as a Codex standard (step 10). The standards acceptable and published as Codex Alimentarius standards have legal binding in nations operating them. As on date, there is no internationally agreed microbiological standards for fishery products.

Codex standards on Fish and Fishery Products are:

1. Canned Crab Meat (CXS 90-1981)
2. Canned Fish (CXS 90-1981)
3. Canned Finfish (CXS 119-1981)
4. Canned Tuna and Bonito (CXS 70-1981)
5. Canned Salmon (CXS 3-1981)
6. Canned Shrimps or Prawns (CXS 37-1981)
7. Canned Sardines and Sardine-Type Products (CXS 94-1981)
8. Quick Frozen Finfish, Eviscerated or Uneviscerated (CXS 36-1981)
9. Quick Frozen Fish Sticks (Fish Fingers), Fish Portions and Fish Fillets – Breaded or in Batter (CXS 166-1989)
10. Quick Frozen Fish Fillets (CXS 190-1995)
11. Quick Frozen Blocks of Fish Fillets, Minced Fish Flesh and Mixtures of Fish Fillets and Minced Fish Flesh (CXS 165-1989)
12. Quick Frozen Shrimp or Prawns (CXS 92 -1981)
13. Quick Frozen Raw Squid (CXS 191-1995)
14. Quick Frozen Lobsters (CXS 95-1981)
15. Salted Fish and Dried Salted Fish of the Gadidae Family of Fishes (CXS 167-1989)

16. Smoked Fish, Smoke Flavoured Fish and Smoke-Dried Fish (CXS 311-2013)
17. Crackers from Marine and Freshwater Fish, Crustaceans and Molluscan Shellfish (CXS 222-2001)
18. Dried Shark Fins (CXS 189-1993)
19. Fish Sauce (CXS 302-2011)
20. Fish Oils (CXS 329 -2017)

8.3.1.1.2. Codex Guidelines for Fish and Fishery Products

There are some guidelines published by Codex for fish and fishery products and there are given as follows:

1. Guidelines for the Sensory Evaluation of Fish and Shellfish in Laboratories (CAC/GL 31-1999)
2. Guidelines on the Application of General Principles of Food Hygiene to the Control of Pathogenic Vibrio Species in Seafood (CAC/GL 73-2010)
3. Guidelines on the Application of General Principles of Food Hygiene to the Control of Viruses in Food (CAC/GL 79-2012)
4. Model Certificate for Fish and Fishery Products (CAC/GL 48-2004)
5. Guideline Procedures for the Visual Inspection of Lots of Canned Foods for Unacceptable Defects (CAC/GL 17-1993)
6. Guidelines on Good Laboratory Practice in Pesticide Residue Analysis (CAC/GL 40-1993)
7. General guidelines on sampling (CAC/GL 50-2004)
8. Guidelines on the Use of Mass Spectrometry (MS) for Identification, Confirmation and Quantitative Determination of Residues (CAC/GL 56-2005)

8.3.1.1.3. FAO Codes of Practice

In addition to the Codex Alimentarius Recommended International Standards for fishery products, a set of Codes of Practice is compiled by the FAO Department of Fisheries. The Fisheries Products and Marketing division of the FAO prepare them in consultation with international experts. The Codes of Practice are voluntary in nature to provide technical guidance to seafood processors to make products to meet Codex Alimentarius Standards. The Codes of Practice are prepared for Fish and Fishery Products (CAC/RCP 52-2003).

1. Fresh fish (FAO Fisheries Circular 318)
2. Frozen fish (FAO Fisheries Circular 145)

3. Smoked fish (FAO Fisheries Circular 321)
4. Canned fishery products (FAO Fisheries Circular 315)
5. Salted fish (FAO Fisheries Circular 336)

The Codes of Practice Fish and Fishery Products Second edition was published in 2012 by WHO/FAO (CAC/RCP 52-2003). It consists of prerequisite programme, general consideration for the handling of fresh fish, shell fish and other aquatic invertebrates, HAAP and defect action point analysis, aquaculture production, processing of live an raw bivalve molluscs, processing of fresh, frozen and minced fish, processing of frozen surimi, processing of quick frozen coated fish products, processing of salted and dried salted fish, smoked fish, smoke flavoured fish and smoke dried fish, processing of lobsters, processing of crabs, processing of shrimps and prawns, processing of cephalopods, processing of canned fish, shell fish and other aquatic invertebrates, transportation and retail.

8.3.1.2. International Organization for Standardization (ISO)

International Organization for Standardization (ISO) is an independent, non-governmental organization (NGO) established in 1947. It is a worldwide federation of national standards bodies from about 130 countries. The mission of ISO is to promote the development of standardization and related activities in the world, with an aim to facilitate the international exchange of goods and services, and to develop cooperation in the spheres of intellectual, scientific, technological and economic activity. The ISO brings in experts through its members to develop voluntary, consensus-based, market relevant international standards to support innovation and provide solutions to global challenges. The ISO 9000 is the most popular standard on quality management is the general series. The countries individually decide whether to mandate adoption of ISO's for agricultural food products. Each national governing body has become a member body of ISO.

8.3.1.2.1. ISO 9000 Series of Quality Management and Assurance Standards

The first edition was published in 1987 based mainly on UK BS-5750 standards, and Canadian standards, CSA-Z299. The ISO 9000 series is adopted to comply with customers requiring standards, to sell the produce in EU market, to compete in domestic market, to improve the quality system, to prevent repetitive auditing and to improve sub-contractors' performance. The ISO 9000 series consists of :

- ISO 9000 (a guide)
- ISO 9001 (a set of requirements for the quality system of the supplier)

- ISO 9002 (products standards)
- ISO 9003 (final inspection and testing)
- ISO 9004 (guidelines for developing and implementing quality system principles, structure, auditing and review)

The benefits of ISO 9000 certification are it helps to improve processes and procedures, to understand processes, to improve existing quality policies and programs, to compile quality manual, to communicate between management and employees, and to improve company's credibility.

8.3.1.2.2. ISO 22000: 2018 Food Safety Management Systems

The ISO 22000:2018 is a requirement for any organization in the food chain, as it sets the requirements for a food safety management system (FSMS). The ISO 22000 is prepared based on the Codex principles for Food Hygiene. It defines the requirements of an organization to demonstrate its ability to control food safety hazards and to ensure that food is safe for consumption. It enables the organization to implement FSMS in place in order to improve their overall performance in food safety. The advantages of ISO 22000 are it provide consistent and safe food products and services meeting regulatory requirements, to improve risk management in food safety processes, and to demonstrate links with the Codex safety guidelines.

The ISO has expanded its wings on agribusiness and agriculture supply chains with the growth in international trade. For example, the internationally standardized freight container enables all components of a transport system such as air and seaport facilities, railways, highways, and even packages to interface among themselves efficiently. The standardized transport system combined with standardized documents helps to identify sensitive or dangerous cargoes and makes the international trade cheaper, faster and safer.

8.3.1.2.3. ISO Standards for Seafood

The ISO has developed six standards based on the TraceFish standards developed under an EU project, which are a statutory requirement in some Asian countries. The ISO standards are formulated for captured and farmed fish, crustacean and molluscan distribution chains. The capture distribution chain standards specify the way traded fishery products are to be identified, the information to be generated, and held on those products by each of the food businesses through distribution chains. The farmed fish distribution chain standards specify the distribution for human consumption the farmed fish and their products, from fish meal, breeding, and farming through to retailers or caterers. The distribution chain includes farming – broodstock collection, hatcheries and nurseries, farms, harvesting; processors; traders and wholesalers; retailers and caterers; logistics including materials brought from other domains; feed production.

1. ISO12875:2011 – Traceability of finfish products – Specification on the information to be recorded in capture finfish distribution chains
2. ISO 12877:2011 – Traceability of finfish products – Specification on the information to be recorded in farmed finfish distribution chains
3. ISO18537:2015 – Traceability of crustacean products – Specification on the information to be recorded in captured crustacean distribution chains
4. ISO16741:2015 – Traceability of crustacean products – Specification on the information to be recorded in farmed crustacean distribution chains
5. ISO18539:2015 – Traceability of molluscan products – Specification on the information to be recorded in captured molluscan distribution chains
6. ISO18538:2015 – Traceability of molluscan products – Specification on the information to be recorded in farmed molluscan distribution chains
7. ISO/PRF 22948 – Carbon footprint for seafood (under development)
8. ISO/PRF 23855 – Frozen surimi- Specification (under development)

8.3.2. National Standards

All national Governments are responsible for public health associated with the consumption of fish and fishery products. They formulate the national or local food laws. These laws are enforced by a team of official inspectors, who have the responsibilities to ensure the absence of parasites, certain chemicals or pathogens in fish products, to enforce to hygiene surveillance, and to ensure operation of fair-trade practices with correct labelling, weights, descriptions, etc.

8.3.2.1. Bureau of Indian Standards (BIS)

Bureau of Indian Standards (BIS) is functioning under the Ministry of Consumer Affairs, Food and Public Distribution, Government of India. It came into existence on 01 April 1987 through an Act of Parliament on 26 November 1986. It was functioning previously as Indian Standard Institution (ISI), which was established on 06 January 1947. BIS has formulated 64 standards related to fish and fishery products, out of which 33 are active. All these standards are voluntary, which addresses method of production, quality and safety requirements. It also stipulates the method of testing and sampling. The FSSAI is planning to re-draft all BIS standards related to fish and fishery products as most of the food safety requirements are not in sync with the current national standards (Table 8.2).

Table 8.2. BIS Standards for Fish and Fishery Products

S.No	IS No.	Year	Standards
1	IS 2168	1971	Pomfret Canned in Oil
2	IS 2236	1968	Prawns/Shrimp Canned in Brine
3	IS 2237	1997	Prawns (Shrimps) - Frozen
4	IS 3336	1965	Shark Liver Oil for Veterinary Use
5	IS 3892	1975	Frozen Lobster Tails
6	IS 4304	1976	Tuna Canned in Oil
7	IS 4780	1978	Pomfret, Fresh
8	IS 4793	1997	Whole Pomfret - Frozen
9	IS 5734	1970	Sardine Oil
10	IS 6121	1985	*Lactarius* sp Canned in Oil
11	IS 6122	1997	*Seer Fish (Scomberomorus spp.)* - Frozen
12	IS 6123	1971	*Seer Fish (Scomberomorus spp.)* - Fresh
13	IS 7143	1973	Crab Meat Canned in Brine
14	IS 7313	1974	Glossary of Important Fish Species of India
15	IS 7582	1975	Crab Meat, Solid Packed
16	IS 8076	2000	Frozen Cuttlefish and Squid
17	IS 9808	1981	Fish Protein Concentrate
18	IS 10059	1981	Edible Fish Powder
19	IS 10760	1983	Mussels Canned in Oil
20	IS 10762	1983	Tuna Canned in Curry
21	IS 10763	1983	Frozen Minced Fish Meat
22	IS 11427	2001	Fish and Fisheries Products - Sampling
23	IS14513	1998	Beche-de-mer
24	IS 14514	1998	Clam Meat – Frozen
25	IS 14515	1998	Fish Pickles
26	IS 14516	1998	Cured fish and fisheries products - Processing and storage-Code of Practice
27	IS 14517	1998	Fish Processing Industry - Water and Ice – Technical ReRuirements
28	IS 14520	1998	Fish Industry - Operational Cleanliness and layout of market - Guidelines (Amalgamated Revision of IS 5735, 581 and 8082)
29	IS 14890	2001	Sardines - Fresh, Frozen and Canned (Amalgamated revision of IS 2421, 6677,8652,8653, 9750 and 10761)
30	IS 4891	2001	Mackerel - Fresh, Frozen and Canned (Amalgamated Revision of IS 2420, 3849,6032, 6033 and 9312)
31	IS 14892	2000	Threadfin - Fresh and Frozen
32	IS 14949	2001	Accelerated Freeze Dried Prawns (Shrimps) (Amalgamated Revision of IS 4781 and 4796)
33	IS 14950	2001	Fish - Dried and Dry-Salted

8.3.2.2. FSSAI Standards for Fish and Fishery Products

The Government of India published notification for FSS (Food Products Standard and Food Additives) 11th Amendment Regulations relating to Vertical Standards for Fish and Fish Products on 17 January 2018. This regulation is applicable for frozen fish, canned fish, smoked fish products, fin fish, crustaceans, mollusks and fish fillets. It also notifies the conditions of freezing process, requirements, potential contaminants, food additives, hygiene, toxins and residues, packaging and labelling regulations for the final fish products. The regulations also include products such as ready-to-eat fish or shellfish curry, sardine oil, edible fish powder, fish pickles, frozen minced fish meat, freeze dried prawns and clam meat.

The FSSAI has notified the Microbiological Standards for Fish and Fish Products through its notification Food Safety and Standards (Food Product Standards and Food Additives) Third Amendment Regulation, 2017 and the regulations came into force on 1st January 2018. The microbiological requirements for fish and fishery products on hygiene indicator organisms with sampling plan and limits for aerobic plate count, coagulase positive Staphylococci and yeast and mold count are given in Table 8.3, along with stages where these criteria apply and the actions required, if results are unsatisfactory. The microbiological requirements for fish and fishery products on safety indicator organisms such as *Escherichia coli, Salmonella, Vibrio cholerae* (O1 and O139), *Listeria monocytogenes* and *Clostridium botulinum* with sampling plan and limits are given in Table 8.4. The microbiological standards are applicable for chilled/ frozen finfish, crustacean, cephalopods, bivalves; live bivalve mollusks; frozen cooked crustaceans and heat shucked mollusks; dried or salted and dried fishery products; thermally processed fishery products; fermented fishery products; smoked fishery products; fish mince/surimi and analogues; fish pickle; battered and breaded fishery products; convenience fishery products; and powdered fish-based products.

Table 8.3. Microbiological Requirements for Fish and Fishery products - Hygiene Indicator Organisms

Product category	**Aerobic Plate Count**	**Coagulase positive Staphylococci**	**Yeast and mold count**
Chilled/ Frozen finfish	$m=5x10^5$ $M=1x10^7$	-	-
Chilled/ Frozen Crustaceans	$m=1x10^6$ $M=1x10^7$	-	-
Chilled/ Frozen Cephalopods	$m=5x10^5$ $M=1x10^6$	-	-
Chilled/ Frozen bivalve	$m=5x10^5$ $M=1x10^6$	-	-
Frozen cooked crustaceans/ Frozen heat shucked mollusk	$m=5x10^5$ $M=1x10^6$	$m=1x10^2$ $M=1x10^3$	-

Dried/ Salted and Dried Fishery Products	$1x10^5$	-	m=100 M=500
Thermally processed fishery products	Commercial sterility	-	-
Fermented Fishery Products	-	$m=1x10^2$ $M=1x10^3$	100
Smoked Fishery Products	$1x10^5$	$m=1x10^2$ $M=1x10^3$	-
Accelerated Freeze Dried Products	$1x10^5$	100	-
Fish Mince/Surimi and analogues	$m=5x10^5$ $M=1x10^6$	$m=1x10^2$ $M=1x10^3$	-
Fish Pickle	$1x10^3$	$m=1x10^2$ $M=1x10^3$	100
Battered and Breaded fishery products	$m=5x10^5$ $M=1x10^7$	$m=1x10^2$ $M=1x10^3$	100
Convenience fishery products	$m=5x10^3$ $M=1x10^4$	$m=1x10^2$ $M=1x10^3$	-
Powdered fish-based products	$m=5x10^4$ $M=1x10^5$	m=1x10 $M=1x10^2$	100

Table 8.4. Microbiological Requirements for Fish and Fishery products - Safety indicator organisms

Product category	***Escherichia coli***	***Salmonella***	***Vibrio cholera* (O1 & O139)**	***Listeria mono-cytogenes***	***Clostridium botulinum***
Chilled/ Frozen finfish	m=11 M=500	Absent/25g	Absent/25g	-	-
Chilled/ Frozen Crustaceans	m=11 M=500	Absent/25g	Absent/25g	-	-
Chilled/ Frozen Cephalopods	20	Absent/25g	Absent/25g	-	-
Live bivalve mollusk	m=230/100g M=700/100g	-	-	-	-
Chilled/ Frozen bivalve	46	Absent/25g	Absent/25g	-	-
Frozen cooked crustaceans/ Frozen heat shucked mollusk	m=1 M=10	Absent/25g	Absent/25g	Absent/25g	-
Dried/ Salted and Dried Fishery Products	20	Absent/25g	-	-	-
Thermally processed fishery products	-	-	-	-	Absence of viable spores or vegetative cells of *Clostridium botulinum* and absence of botulinum toxin

Fermented Fishery Products	m=4 M=40	Absent/25g	-	-	Absence of viable spores or vegetative cells of *Clostridium botulinum* and absence of botulinum toxin
Smoked Fishery Products	m=11 M=500	Absent/25g	Absent/25g	-	-
Accelerated Freeze Dried Products	20	Absent/25g	Absent/25g	-	-
Fish Mince/Surimi and analogues	20	Absent/25g	Absent/25g	-	-
Fish Pickle	20	Absent/25g	-	-	-
Battered and Breaded fishery products	m=11 M=500	Absent/25g	Absent/25g	-	-
Convenience fishery products	m=1 M=10	Absent/25g	Absent/25g	-	-
Powdered fish-based products	-	Absent/25g	-	-	-

8.4. Assessment of Food Safety Management Programs

Food safety management followed in both the food industry and the government agencies. The responsibility of ensuring food safety lies with the industry with the advances in HACCP system. The industry must provide evidence that it is aware of the risk associated with its products and preventive measures to control the risks. The establishment of food safety laws, standards and regulations are mainly to verify whether is industry is complying with them. It is the role of regulatory authorities is to verify it through assessment.

The term "assessment" refers to an industry or governmental activity to verify that the food safety management system is implemented correctly and effectively, and is maintained. It is part of the verification in a HACCP system. In industry, the term used is "**audits**". There are variety of ways it is used such as audit of systems, processes and procedures, projects, laboratories, manufacturing, organizations and their management.

In the framework of the enforcement, governmental authorities verify the compliance of industry practices with laws and regulatory requirements. This activity is referred to as "**inspection**". In earlier days, inspections are a snap-shot visit for checking compliance with good hygienic practice (GHP). With advances of HACCP and the development of an integrated approach to food safety management, various procedures and scope for inspection are evolved, which is now referred as "**regulatory audits**". Regulatory assessment is to check whether control/preventive measures are implemented correctly by the industry, however, government inspection cannot be a measure to ensure safety.

Auditing processes are used for the evaluation of the capability of a supplier to provide a raw material according to given safety specifications, evaluation of equivalence in control measures in case of export–import certification, or evaluation of the status of a factory or a business, other than for verifying compliance.

The different types of food safety assessments are:

1. Internal assessments carried out by industry (part of self-control);
2. External assessments carried out by either
    a. Regulatory agencies (i.e. inspection); or
    b. Third party assessments by customers or certification bodies.

References

Connell, J.J. 1995. Control of fish quality. Wiley-Blackwell; 4th Edition, 256 pp.

CAC 2001. Food Hygiene. Basic Texts. 2nd ed. Codex Alimentarius Commission of Food and Agriculture Organization / World Health Organization, Rome, Italy.

EC 1991. Council Directive 91/493/EEC of 22 July laying down the health conditions for the production and placing on the market of fishery products. Official Journal of the European Commission. No. L268. pp.15-34.

EC 1994. Commission Decision of 20 May 1994 laying down detailed rules for the application of Council Directive 91/493/ EEC, as regards own health checks on fishery products. Official Journal of the European Commission No. L 156. pp. 50-57.

EC 2000. Proposal for a Regulation of the European Parliament and of the Council on the Hygiene of Foodstuffs, European Commission, Brussels, Belgium. pp.17-42.

FAO 1995. Code of Conduct for Responsible Fisheries. Food and Agriculture Organization, Rome.

FAO 1998. FAO Technical Guidelines for Responsible Fisheries. No. 7. Food and Agriculture Organization, Rome.

FDA 1995. Procedures for the Safe and Sanitary Processing and Importing of Fish and Fishery Products- Final Rule. Code of Federal Regulations, Parts 123 and 1240. Volume 60, No 242, 65095-65202, Food and Drug Administration, USA.

9

Legislation for Quality Assurance in Exportable Indian Seafood

Introduction

Export of quality seafood products is important to have a success in international marketing. Exporters must be quality conscious and the Governments must insist on quality control and inspection. The seafood export industry in India is over 50 years old. Seafood contributes 10% of the total exports and nearly 20% of all agricultural exports. In the year 2022-23, India exported 17,35,286 MT seafood to more than 100 countries worth US $8.09 billion equivalent to Rs. 63,969.14 crore. There are over 150 fish processing facilities approved by the European Union out of the existing 350 processing facilities. The Export Inspection Council and the Marine Products Export Development Authority are working very closely with the fish processing industries for the growth of seafood exports.

The Government of India enacted the **Export (Quality Control and Inspection) Act, 1963** for the development of export trade through quality control and inspection. It empowers the Government to notify the products intended for compulsory quality control and inspection prior to their export. It specifies the type of quality control required for the notified products. It recognizes standard specifications for a notified product. It prohibits the export of notified products that are not complying with export conditions. It creates a machinery for quality control and take measures to protect markets for Indian commodities. As the first step, **Export Inspection Council** (the Council) was established on 1st January 1964 for implementation of various provisions of the Export (Quality Control & Inspection) Act, 1963.

9.1. Export Inspection Council (EIC)

The Export Inspection Council (EIC) is the first official Export-Certification Authority, formed to certify the safety and quality of products exported from India. It is established by the Government of India under section 3 of the Export (Quality Control and Inspection) Act, 1963. The EIC functions under the Ministry of Commerce & Industry of the Government of India. The

EIC is the only regulatory organization in India having global acceptance. It has constituted specialist committees such as Administrative Committee, Technical Committee and Standing Committees. The EIC either directly or through Export Inspection Agencies (EIAs), its field organizations render the services. The EIC provides mandatory certification for various food items such as fish & fishery products, dairy products, honey, egg products, meat & meat products, poultry meat products, animal casing, gelatine, ossein, and crushed bones and feeds additive; and pre-mixtures.

9.1.1. Vision and Mission of EIC

The Vision of EIC to facilitate worldwide access for Indian exports through a credible and efficient inspection and certification system and to earn global recognition as India's premier organization for certifying quality.

The Mission of EIC is

- To create an export inspection & certification infrastructure within the country based on International Standards for Certification Authorities in consonance with WTO requirements
- To instill confidence in importers about quality and safety of Indian exports
- To provide accredited state-of-art testing facilities in frontier areas
- To enhance capability of manpower through trainings to meet International requirements
- To obtain recognition of India's export certification system from our major trading partners
- To participate in International fora and project Indian interest
- To be in sync with the latest technological advancements for capacity building

9.1.2. Role of EIC

The EIC assures the quality and safety of the products as specified in the Act by conducting either quality assurance or food safety management based certification or a consignment wise inspection. The major roles of EIC are

1. To provide certification of quality of export commodities through installation of quality assurance systems (In-process Quality Control and Self-Certification) in the exporting units as well as consignment-wise inspection.
2. To issue certification of quality of food items for export through installation of Food Safety Management Systems in the food processing units as per international standards.

3. To issue different types of Certificates such as Health, Authenticity etc. to exporters under various product schemes for export.
4. To perform laboratory testing services.
5. To provide training and technical assistance to the industry in installation of Quality and Safety Management Systems based on principles of Hazard Analysis Critical Control Point (HACCP), ISO-9001: 2000, ISO: 17025 and other related international standards, laboratory testing etc.
6. To recognize Inspection Agencies as per ISO 17020 and Laboratories as per ISO 17025 and utilizing them for export inspection and testing, respectively.

9.1.3. Export Inspection Agency (EIA)

The Export Inspections Agencies are established in 1966 under section 7 of the Quality Control and Inspection Act 1963. The EIA is a field organization of EIC for implementing the measures, standards and polices. Five EIAs are established at Mumbai, Kolkata, Kochi, Delhi and Chennai for quality control and inspection of products as notified under the Act. The EIAs operate under the technical and administrative control of EIC. The EIAs are intended for export of products from various States and Union Territories. There are 30 sub-offices and State-of-art NABL accredited laboratories created later to cater to the needs of exporters located at all major ports and industrial centres.

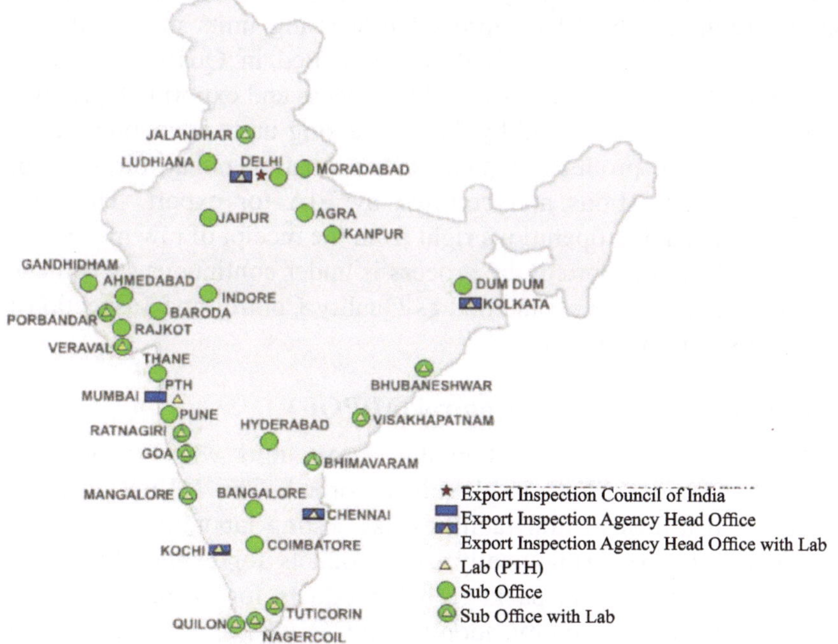

9.1.4. Quality Control and Inspection of Fish and Fishery Products

The ISO defines "Quality Control" as the operational techniques and activities used to determine the quality requirements of a commodity to satisfy the standard specifications. "Inspection" means the process of determining whether a batch of products in that commodity complies with the standard specifications and any other specifications stipulated in the export contract. The EIA is responsible for quality control and pre-shipment inspection of fish and fishery products. The following schemes are followed for fish and fishery products.

A. Consignment wise inspection (End product inspection)

The consignment wise inspection involves inspection of final products and hence termed as end product inspection. Samples are drawn and subjected to detailed inspection to ensure conformity with the prescribed standard specifications. Certain recognized inspection agencies inspect the export consignment and issue certificates. This approach of quality control was only in force at the time of establishment of EIA. Small scale manufacturers make use of consignment-wise inspections. Now-a-days, dried fishery products are inspected by this system.

B. In-process quality control (IPQC)

The IPQC scheme is introduced by the Government of India in 1973 for fish and fishery products. The EIA approved processing units must fulfill the minimum sanitary and hygienic facilities stipulated in Quality Control & Inspection rules. Such units are permitted to process and export fish products. The quality inspection conducted by fish processing units is counterchecked by the EIA. The final products is subjected to organoleptical, physical and bacteriological examinations and certified by EIA for export. The entire processing and packaging operations right from the receipt of raw material till the finished product is exported, the process is under continuous surveillance by EIA. The scheme is later remained as Quality Control and Inspection in Approved Units (QCIA) in 1987.

C. Modified In-Process Quality Control (MIPQC)

The MIPQC system is introduced in the export units where approval to process products under IPQC is already accorded. The MIPQC provides certain additional infrastructural facilities like testing laboratory to produce organoleptically and bacteriologically safe products for export. Such units are permitted to process and package fish products for export under their own supervision and control by adopting IPQC measures, in addition to organoleptic and bacteriological examination of final products. The EIA assists and guides the export units for producing quality products. The export units are responsible for ensuring the quality and safety of the product meant for export, but the inspection certificate is issued by the EIA.

D. Self-certification

In self - certification scheme, the export units are responsible for ensuring the quality of their products as required by law. The export units act as Inspection Agency to certify their own product for export. The export units that have performed well without rejection under IPQC scheme for a stipulated period are only eligible for self-certification scheme. The export units must have the required infrastructure facilities, strict quality control system, and regular quality auditing program to become eligible for self-certification scheme.

E. Voluntary Inspection

The voluntary pre-shipment inspection is performed for the benefit of foreign buyers in respect of products that are not brought under compulsory pre-shipment inspection.

F. EU Approved Unit Inspection

The seafood export not meeting the safety requirements with respect to pathogens suffer ban on import by European Union (EU). The Government of India adopted a new system for approval of seafood processing units exporting to EU countries as per Directive 91/493/EEC. In this scheme, all the EU approved units are monitored and inspected by a team of experts (Supervisory Audit Team) once in 2 months. The processing unit must implement SSOP, GMP and HACCP with all relevant documents and have regular monitoring procedures.

9.1.5. Executive Instructions on Fish and Fishery Products

The EIC has given a total of 14 executive instructions to improve the exports from India, which includes basmati rice, black pepper, fish and fishery products, dried fish, dried fish maws, honey, live fish, milk products, eggs products, poultry and poultry meat, raw meat (chilled/frozen), animal casing, crushed bones, ossein and gelatin, and fruit products. The list of executive instructions related to the fish and fishery products is given below:

1. Executive Instructions for Approval and Monitoring of Fish and Fishery products for Exports (Doc. No. EIC/F&FP/Ex.Inst./March/2012/Issue 4)
2. Executive Instructions for Approval and Monitoring of Establishments Processing Live Fish for Exports to the countries other than the European Union (Doc. No. EIC/Live Fish/Ex.Inst.(Non-EU)/March/2012/Issue 2)

3. Executive Instructions for Quality Control and Pre-shipment Inspection of Dried Fish and Dried Fish Maws for Export (Doc. No. EIC/Ex.Inst./Dried fish/July/2003/Issue 1 and Doc. No. EIC/Ex.Inst./Dried fish maws/July/2003/Issue 1)
4. Executive Instructions for Scheme for Approval and Monitoring of Establishments/ Factory vessels/ Freezer vessels Processing/ Storage of Fish & Fishery Products for Export (Doc No. EIC/F&FP/Ex. Inst./August /2005/Issue 3)
5. Executive Instructions on Raw Meat (Chilled/Frozen) Certification (Doc. No. EIC/Raw Chilled/Frozen Meat-Ex. instruction/Dec. 2003/Issue 1)

9.1.6. Scheme for Approval and Monitoring of the Facility Exporting Fish and Fish Products

The Executive Instruction (Doc. No. EIC/F&FP/Ex.Inst./March/2012/Issue 4) provides information on the approval and monitoring of fish and fishery products for exports. The requirements for the approval of the feed mills, hatcheries, aquaculture farms, fishing harbours, landing / auction centres, fishing vessels, factory vessels, freezer vessels, pre-processing centres, ice plants, establishments and cold storages to process or to undertake allied activities are published by several GOI orders and notifications. These facilities have to be approved by the Competent Authority [EIC for EU, Russian Federation (RF) and EIAs for Non-EU countries] on their compliance to the requirements of the Notification No. S.O. 730 (E) dt. 21.08.1995. Minimum requirements for the aforesaid facilities are given in various appendices.

9.1.6.1. Scheme for Approval of the Facility

The facilities have to submit an application for approval in the prescribed format in duplicate along with documents to the nearest office of EIA. Application fee and assessment fee have to be paid in favour of the EIA concerned. The application must contain the following documents:

1. HACCP Manual (including SSOP, GMP, process flow chart (s) with product description)
2. Facilities that export to the EU and RF, certified copy of test report of water complying requirements of Council Directive 98/83/EC dt 3.11.1998; and those export to countries other than EU and RF, those tested as per IS: 4251. Samples can be tested either in EIA laboratories or EIC approved labs. In the case of cold store, water test report is not required.
3. Layout plan covering all sections including machineries (site plan and building plan)

4. Certified copies of legal identity of the applicant and scope of operations.
5. Certified copy of lease agreement for premises and building, if necessary.
6. Certified copy of registration certificate issued by registering authority.
7. Bio-data of the technologist(s) and certificate of approval issued by EIA for approved technologists. In case of hatchery, feed mill, freezer vessel, ice plant and cold store, bio-data of competent personnel.
8. An Undertaking and Guarantee, if applicable.
9. Certified copy of consent to operate letter issued by State Pollution Control Board.
10. Certified copy of the allotment order of Importer Exporter Code number (IEC).

9.1.6.1.1.Processing of Application for Approval

- Applications received are scrutinised within 3 working days by the EIA office and discrepancies, if any are communicated to the applicant for rectification.
- Application is then forwarded to the Head Office of the EIA within 7 working days of receipt.
- HACCP manual and SSOPs are assessed by the authorized EIA officer (s) and an adequacy audit report is given with the recommendations.
- Discrepancies, if are communicated to applicant after scrutiny at the Head Office of the concerned EIA for rectification.
- The complete application along with HACCP document submitted is then forwarded by the EIA to the Convener of Assessment Panel of Experts (APE) for assessment in the applicant's facility within 15 working days from the date of receipt. (In case of the factory vessels, freezer vessels, pre- processing centres, establishments, the APE assess the unit in two stages. In the case of feed mill, hatchery, aquaculture farm, fishing harbour, landing/ auction centre, fishing vessel, ice plant and cold storage, the assessment is done in one stage only and there is no conditional approval)
- In the first visit, the APE assesses the infrastructure and equipment facilities to avoid cross contamination, and also their compliance to regulatory requirements and if satisfied, recommends for the conditional approval. On receipt of the recommendation, the Technical Division of EIC process it and the Director (I&QC) grant conditional approval within 7 working days. Conditional approval is initially granted for a period of 3 months and later can be extended for a period of 6 months. In case over this period, the

conditional approval stands withdrawn and the facility must submit fresh application for approval.

- Once conditionally approved, the facility starts the production of fish and fishery products or carry out related activities. The export to the non EU countries is allowed only after full approval. In case of the EU and the RF, the export is permitted only after enlisting the facility in the approved establishment list available in their website.
- Conditionally approved facility intimates the EIA once production is started. The APE makes its second visit to assess the processing methods which are in progress, and to conduct HACCP audit. If found satisfactory, full approval is granted for a period of 2 years from the date of conditional approval. A certificate of approval is issued by the EIA in case of export to non-EU countries and by the Director (I&QC) of EIC in case of export to the EU and the RF .
- If a non-EU approved facility submits application for the approval to export to the EU or to RF, there is no need for conditional approval. The APE can assess the infrastructure, equipment and HACCP implementation in the first visit and if satisfied, recommend for the full approval.

9.1.6.1.2.Assessment Panel of Experts (APE)

- The APE is constituted by EIC and the experts are nominated by the EIA.
- The APE comprises representatives from EIC, EIA, CIFT, MPEDA, CAA, CMFRI, NFDB, State Fisheries Departments, Fisheries Colleges, Harbour Authorities, Port Trust Authorities, Trade Associations or Empanelled Experts.
- The Convener of the APE must be an EIA representative in the cadre of Deputy Director. In unavoidable circumstances, senior most Assistant Director can be the EIA representative.
- The quorum of APE is two.
- If the APE finds any deficiency, a Non- Conformity Report (NCR) is given for taking corrective action within one month period.
- The Convenor of the APE submit the assessment report to the EIA within 3 working days after completion of the visit. The APE can give a maximum of 3 months time, if verification is needed and submit the report to the EIA within 3 working days of verification.
- The laboratory expert of the APE assesses the in-house laboratory of the facility and submit recommendation for approval of the laboratory.
- In the case of frozen fishery products, the APE assesses the production capacity of the factory vessels / establishment based on the Operational

Freezing Capacity (calculated based on the actual quantity of fish products frozen per day).

- The APE assesses the production capacity of the freezer vessels based on the freezing and frozen storage capacity of the vessel.
- The APE assesses the minimum number of approved technologists/ competent personnel required by the facility.
- The report of the APE visit is examined by the EIA.

9.1.6.1.3. Renewal of Approval of the facility

- The EIA reminds the facility about renewal 90 days before the expiry of current approval.
- Approved facilities seeking renewal must submit application at least 60 days in advance of the expiry of earlier approval to the EIA.
- Application (s) received are scrutinised in a similar manner as done for original approval
- In case the facility does not apply for renewal in time within the validity period, the approval granted lapses and the unit has to apply for fresh approval.
- The APE assesses the facilities, hygiene & sanitation, own-check system, implementation of statutory requirements, maintenance of records etc. to ensure proper maintenance and adequacy of control measures, and also conduct HACCP audit.
- If the APE recommends, the EIA grant renewal of approval for a period of two years from the date of expiry of earlier approval and issue the Certificate of Approval.

9.1.6.2. Scheme for Monitoring the Facility

- The EIA officials carry out periodic monitoring of the approved facility as per the regulatory requirements and those specified by the EU / importing countries are compliance.
- Monitoring is done by an EIA officer of the level of Assistant Director / Technical Officer.
- Monitoring officials verify the own checks system adopted by the facility.

9.1.6.2.1. Frequency of Monitoring

a. Monitoring visits are carried out **once in a month** after approval of the pre-processing centres, ice plants, establishments and cold storages. If the performance is satisfactory for a year without any foreign rejection or complaint, the frequency is reduced to once in **two months**. If the performance is satisfactory for another one year, the frequency is reduced to once in **three months**. In case, the performance is non-satisfactory, the frequency is reverted back to once in a month and the process is repeated.

b. Monitoring visits are carried out at a frequency of **once in a month** for factory vessels / freezer vessels are monitored initially. If it is not possible, it is done **during each berthing**. If the performance is satisfactory, the frequency can be re-fixed.

c. Monitoring visits are carried out once in **six months** on initial approval of feed mill, hatchery, aquaculture farm, fishing vessels and fishing harbours/ landing centres. If the performance is satisfactory for one year, frequency is re-fixed to **once in a year**. If the performance is unsatisfactory, frequency is reverted back to six months.

9.1.6.2.2.Areas of Monitoring

1. **Facility checks**: to ensure that all approved facilities are well maintained, including sanitary and hygienic conditions.
2. **HACCP Implementation**: to ensure the implementation of HACCP as envisaged in HACCP manual. It includes verification of CCP, GMP, GHP, SOP, SSOP, traceability, good storage practices, raw material / process/ product controls, time / temperature controls, controls on additives/ preservatives, quality management of water and ice, calibration and validation, etc. In case of aquaculture products, to ensure that raw material is procured from registered aquaculture farms approved by EIAs.
3. **Testing and laboratory practices**: to ensure that the sampling procedures and test methods used are adequate and reliable, including GLP.
4. **Records**: to ensure that the records are maintained in order and address all the controls. In case the facility pays monitoring fee based on F.O.B. value of its exports, random verification of export documents including shipping bill and freight memo. Pre- harvest test certificates is verified.
5. **Fraud control**: to ensure that the facility is not violating the laid down norms, especially in procurement of fish products pre-processed in unauthorised establishments, export of products processed in un-authorised premises, storage of products in unauthorised storages, storage of fish products of other facilities without prior permission of EIA, misuse of CFE (certified fraud examiner) blanks, improper labelling, excess production, etc. The **"Illegal soaking"** of fish products for weight gain is also verified.

9.1.6.2.3. Drawal of Official Samples

To ensure the wholesomeness of the products and effectiveness of cleaning and sanitization operations., samples are drawn for testing sanitation & hygiene control, microbial and residual parameters. Samples for **organoleptic checks** is drawn at random from the raw material, pre- processed materials and processed materials before freezing or packing. The details are recorded in the relevant registers of the facility and in the monitoring report. Defective lot is not allowed for further processing.

9.1.6.2.4. Additional Checks

- Chlorination levels of water used for processing, ice manufacture, foot dip, hand dip, washing utensils / tables etc. A general guideline for chlorination is given below:

1	Processing	2 ppm. or below.
2	Glazing	2 ppm. or below.
3	Ice manufacture	2 ppm. or below.
4	Hand sanitization dip	20 ppm.
5	Foot sanitization dip	50-100 ppm
6	Washing of tables, equipment and utensils	50-100 ppm.
7	Washing of floor	100-200 ppm.

- Temperature of products during receipt, processing and storage.
- Temperature of chill room(s), cold storage, cooker / blancher, chillers etc.
- Belt speed, time taken for cooking / freezing etc.
- Validation of cooking / process / equipment etc., as applicable.
- Rinsing of food contact surface with potable water after every sanitization process.

9.1.6.2.5. Parasite Checks

The monitoring officials conduct visual checks on the available raw material for the presence of visible parasites. Samples are selected from different sources of raw materials available. Parasite infested materials are not be allowed for further processing. The observations are recorded in the raw material register and in the monitoring report.

9.1.6.2.6. Microbiological / Chemical Checks

The monitoring officials draw samples for testing microbiological and chemical parameters, the frequency of checks is given below:

S.No.	Parameters	Products/ Stage	Frequency
1	TPC, *E.coli*, Coagulase positive *Staphylococcus, Salmonella, V. cholerae, V. parahaemolyticus*	Raw / process material	Every monitoring visit
2	TPC, *E.coli*, Coagulase positive *Staphylococcus, Salmonella, V. cholerae, V. parahaemolyticus*	Frozen/ chilled Finished products	Every monitoring visit
3	TPC, *E.coli*, Coagulase positive *staphylococcus, Salmonella, V. cholerae, Listeria monocytogenes*	Frozen cooked crustaceans	Every monitoring visit
4	TPC, Coliforms, *V. cholerae*	Water	Every monitoring visit
5	TPC, Coliforms, *V. cholerae*	Ice	Every monitoring visit
6	TPC, Coliforms	Swabs from food contact surfaces	Every monitoring visit
7	*V. cholerae*	Swabs from worker's hand	Every monitoring visit
8	Antibiotics (Chloramphenicol, Metabolites of Nitrofuran, Tetracycline, Oxy tetracycline, Chlortetracycline)	Raw material / finished products	Once in six months
9	Cadmium, Lead, Mercury,	Raw material / finished fishery products	Once in six months
10	Pesticides (Organochlorine compounds)	Raw material/ finished aquaculture products	-do-
11	Histamine	Raw material/ finished products	-do-
12	TVB-N & TMA-N (If doubt on organoleptic examination)	Raw Material/ Pre- processed Material/ Processed Material	Every monitoring visit.
13	Sulphite & added Phosphates for crustaceans	Raw material/ finished products	Once in six months
14	PSP/DSP in Molluscs (Gastropodas, Bivalvia, Crabs)	Raw material/ finished products	-do-
15	Coagulase positive *Staphylococcus* & sulphite reducing *Clostridium* and purity test	Salt used for processing	-do-
16	Sterility test for Canned/ Retort pouch products)	Fishery products (Canned / Retort Pouch Packed)	Every monitoring visit
17	PH, Yeast & Mould and salt content	Acidified fishery products	Every monitoring visit

18	Moisture content, salt content and yeast & mould, TPC, *E.coli*, Coagulase positive *Staphylococcus*, *V. cholerae* *Salmonella* & Acid insoluble ash .	Dried / salted & dried fishery products.	Every monitoring visit
19	Parameters for the other commodities like Fish oil	Rancidity / Moisture	Every monitoring visit
20	Dioxin & PCB	Fishery products	Every one year

9.1.6.2.7. Sanitation & Hygiene Control Samples

The monitoring officials draw samples for checking the sanitary conditions and hygienic practices of the facility. The swab samples are drawn either before start of the work or after normal cleaning & sanitization, if processing is in progress.

1.	Water used for processing	1 sample
2.	Ice	1 sample
3.	Swabs of food contact surfaces (1) Table (2) Utensils (3) Freezing Tray (4) Workers hand (5) Workers hand (APW *for V. cholerae*)	 1 sample 1 sample 1 sample 1 sample 1 sample

Maximum Permissible limits

S.No	Samples	TPC at 37°C	TPC at 22°C*	Coliforms	*V. cholerae*
1.	Water	50 per ml**	100 per ml	Nil	Absent
2.	Ice	50 per ml**	100 per ml	Nil	Absent
3.	Table / utensils / trays etc	100 per sq.cm	-	Nil	-
4.	Workers Hand	100 per sq.cm	-	Nil	Absent

* Not required for Non-EU establishment

** For EU establishments, the limits of TPC in water and Ice are 20 per ml.

9.1.8. Interpretation of Test Results

To ascertain the performance of the in-house laboratory, the monitoring officials draw 2 sets of samples from the selected production batch during the monitoring **at least once in six months**. One set of samples is sent to EIA Lab and the other set is given to the in-house laboratory for testing all microbiological parameters. No fee will be charged from the processor.

The facility submits the report to the EIA and the EIA compare the test results of both laboratories. If variation above 10% is observed, the same is communicated to the In-charge, who in turn deputes a suitable officer from the

EIA lab to audit the in-house lab. The copy of audit report is then sent to EIC for information.

In case of minor non-conformities, the facility undertakes immediate corrective action and the same is verified by the EIA in the subsequent monitoring visit. In case of major non conformities, the facility must stop the analysis in the in-house lab and send the own check microbiological samples to the EIA lab for testing, till the non-conformities are rectified and the same is verified by the EIA lab expert.

9.1.8.1. Reporting System

- After completing the monitoring, a report is prepared along with the Non Conformity Report (NCR) and suggestions for improvement. The report is submitted to the EIA within 3 working days of the visit. Test reports can be given to the processor, if requested.
- Non-conformities observed during the surveillance visits are recorded in the NCR and a copy provided to the facility for taking corrective action of deficiencies within an agreed time period.
- In case any major discrepancies affecting the food safety, the same is brought to the notice Deputy Director immediately for timely action.

9.1.9. Complaint Handling Procedures

When a complaint is received from the importing country or a consignment of fish products is detained or specific control measures are imposed by the importing countries on food safety or quality grounds, such as product contamination with microorganisms or with harmful residues or due to other reasons including spoilage, incorrect labelling, temperature abuse etc, the EIC/EIA is the Competent Authority to take action as per the complaint handling procedure. Action is required for **Alert Notifications** and **Border Rejection Notifications** issued by the European Commission (EC) under Rapid Alert System for Food and Feed (RASFF). Any **"Information Notification"** or **"News"** received from the EC under RASFF does not warrant official action, however communicated to the approved facility by the EIA for information, necessary action and comments.

- The EIC scrutinizes the complaint received from the foreign countries and informs the EIA concerned immediately for taking action.
- If a processor receives any information regarding rejection of its products at the destination before the Competent Authority receives the same, the processor must inform the EIA concerned giving all details. The EIA shall communicate the same to EIC immediately.

- The EIC can seek clarification from the importing country / Health Authorities, if required.

9.1.10. Action by EIAs

9.1.10.1. Placing the Facility Under "Internal Alert"

The EIA place the processing facility under **'Internal Alert'** immediately after the receipt of the information about the rejection and collect the information from the processor based on the nature of rejection or complaint. It includes full particulars of the consignment, pre-harvest test report, test reports of raw materials and finished products, pre-export test report, stock held by processor, details of investigation, confirmatory test or survey report taken in detained country, and any other relevant information.

9.1.10.2. Information to the Laboratories

Every laboratory involved with the product by way of sampling and / or testing is informed by EIAs about the complaint with a request to investigate the root cause for non-detection of the contaminant and to send a detailed report proposing corrective actions and measures to prevent the recurrence. The EIA examine the report of root cause analysis done by the laboratory and forward the same to EIC with its comments.

9.1.10.3. Enhanced Monitoring Visits

In case of rejections due to quality and / or food safety issues, the frequency of monitoring visit is increased to **2 visits per month**, with special focus on the factor(s) that caused the rejection. Fee per visit for the monitoring visits is borne by the processor. The increased frequency is discontinued, if **four** monitoring visit reports and test reports are satisfactory. In case of unsatisfactory performance, the same frequency is continued till two consecutive satisfactory conditions are achieved.

9.1.10.4. Testing of Consignments

In case of rejections from **EU on microbiological grounds**, the **next 5 consecutive consignments** meant for export to the EU and **one in two consignments** meant for non-EU are tested by EIA for the specific contaminant(s) at EIA Lab. For testing, each batch is treated as a lot and a composite sample is collected from each batch following the sampling scale $(\sqrt{n}+1)/2=x$ (where n is the number of cartons in the lot; x is the number of cartons to be selected, from which a composite sample is collected). Only the batches found satisfactory are allowed for export. The frequency of inspection is till such time 5 consecutive consignments to EU or 2 to non-

EU are cleared. If any EU consignment fails during testing of 5 consecutive consignments, two more consignments are tested until they get cleared. If any non-EU consignment fails, testing continue till two consecutive consignments get cleared after testing.

In case of any complaint received from **non-EU countries on microbiological grounds,** the **next 5 consecutive** consignments are tested for the specific contaminant by EIA prior to shipment irrespective of destination.

In case of rejection due to **residues, 3 consecutive batches** are tested with specific reference to the residue(s) by EIA. A composite sample each from 3 different batches as close to the code(s) of the rejected consignment is tested for the specific contaminant (s) by EIA. If any of the batches fail to meet the requirements, **5 more batches** of the same product or source are tested, till such time they get cleared.

In case of EU rejection due to **organoleptic quality defects** such as spoilage, discolouration etc., **next 5 consecutive consignments** meant for export to the EU and **one in two consignments** meant for the non-EU are inspected for organoleptic factors by EIA. In case of doubt, the same is subjected to chemical test for TVB-N/ TMA-N.

In case any consignment is rejected on **grounds of temperature abuse / rupture of cold chain / labelling defect,** the processor concerned is asked by the EIA to explain the reasons for the rejection, to clarify whether the lapse is on the part of the processor and to justify with adequate proof that due diligence is exercised. If the processor is able to prove that the complaint / rejection is on the factor(s) beyond its control after exercising due diligence, the in-charge of the EIA can forward the explanation to EIC with necessary recommendation indicating that no further action is needed. If the EIA is not satisfied with the explanation, the **next 5 consignments** are checked for core temperature of the frozen fishery product or correct labelling following the sampling scale. Only the consignments that meet the requirements are allowed for export. The cold storage temperature is checked to ensure that it is maintained at -18°C or below before drawal of samples.

9.1.11. Assessment of the Approved Facility

- The EIA arrange a detailed assessment of the facility within a week to determine the root cause of rejection/ complaint, to suggest remedial measures and to collect details of the rejected consignment.
- A detailed root cause analysis is done by the APE to ascertain the actual cause of rejection. The audit is carried out to address the specific contaminant. The assessment covers HACCP and pre-requisite programme, control

measures exercised at all stages of production, storage and transportation, source of raw materials / ingredients / additives, traceability system of the unit, testing of raw materials / ingredients / additives, pre- harvest test reports, pre-export test reports, supplier selection and supplier audit, internal audits, training of employees, validation of HACCP plan / validation of critical limits, good laboratory practices, testing of process / finished product samples.

- In case of rejection due to microbiological factors, in-house lab is audited by the lab expert of the APE to verify whether the lab has any major deficiency in equipment, methodology or analyst with specific reference to the specific contaminant. In case of rejection due to residue, the APE assesses the concerned farm, hatchery and or feed mill, landing sites, fishing vessels to find the cause of rejection.
- If the APE finds it necessary, the assessment of sanitation & hygiene control, raw material / process / product / ingredient etc, samples are drawn for testing at EIA lab / EIC approved labs to ascertain the cause of contamination.

Based on the assessment, the APE prepares a detailed objective report and submit to the in-charge of EIA within 3 working days after completion of assessment. The in-charge of the EIA scrutinises the assessment report and test report(s) of samples drawn during assessment and take appropriate actions:

1. In the case of rejections due to food safety / quality issues, if the APE opined that the facility is capable of producing safe / quality fishery products based on satisfactory performance evidenced by proper HACCP implementation and adequacy of the own check system, the EIA forward the assessment report along with test report(s) to EIC for information.
2. In the case of rejection for reasons other than food safety/ quality issues and if the performance of the unit is satisfactory based on the APE report, the EIA propose to the EIC to revoke the **'Internal alert'**. In the case of non-EU units, the in-charge of the EIA can revoke the Internal alert under intimation to the Director.
3. If the APE report indicates satisfactory performance subject to rectification of minor defects to prevent the recurrence of rejection, the EIA communicate the defects for time bound rectification. The corrective actions are then verified by the EIA and if satisfied, take appropriate action as mentioned above.
4. In case the test result of any sample drawn is found not in conformity with the requirements, next 5 consecutive consignments meant for export are subjected to testing / inspection by EIA for the specific contaminant/

quality parameters at EIA lab and the consignments are permitted for export only after satisfactory test / inspection results.

5. In case the investigation team observed the non-conformity in the areas of primary production, then the same is dealt with by taking suitable action on these links.
6. In case the performance is found unsatisfactory during assessment based on improper HACCP implementation and / or inadequate controls to prevent non-conformity / contamination of fishery products processed in the unit, the in-charge of the EIA take the actions with intimation to EIC.

9.1.12. Action in case of Unsatisfactory Assessment Report

1. Processor is required to show cause within 10 days as to why the approval granted to the facility should not be withdrawn for the lapse(s) brought out in the assessment report.
2. Production and export to all countries is stopped till the approved facility rectifies all the deficiencies pointed out by the assessment team and proper steps are taken to prevent further rejection.
3. Once the processor informs the EIA that the corrective actions are carried out, verification, is done by a Deputy Director level officer. The processor is then allowed to resume production and export only, if the in charge of EIA is satisfied.
4. If the in-charge of EIA is not satisfied with the reply of the processor to the show-cause notice, or with the corrective actions, the approval granted to the facility is withdrawn with the approval of the Director (I&Q/C).
5. After resumption of production, an officer, not below the rank of Technical Officer is deputed to such facility for a minimum period of 10 days extendable upto 30 days to continuously monitor the enforcement of various standards relating to quality control, food hygiene and food safety. The cost of such deputation of EIA officers is charged to the facility. If working is more than 1 shift, all shifts should be covered at random.
6. Only after the clearance from the concerned EIA, based on satisfactory test or inspection results as mentioned under monitoring, the consignment produced are allowed for export.

9.1.13. Dealing with Returned Consignments

The processor can request the EIC to issue a letter to the concerned Health Authority of the importing country for bringing back the rejected consignment through the EIA concerned along with full details of rejected consignment, including type of product, quantity, FOB value, container no, Health Certificate no, & date, reason for rejection.

When a consignment is brought back to India after rejection at the importing country, the processor / exporter must inform the local EIA office the date and port of delivery, enabling the EIA to arrange for inspection of the returned cargo and to collect necessary samples for testing. If the port does not have facilities for sampling and inspection, the returned consignment is taken, under EIA escort, to an approved storage where necessary facilities exist, the details are furnished to the EIA by the processor / exporter. The team comprises of an EIA official (Inspection) and one EIA laboratory official (Sampling). Inspection and sampling are done by the team at the unloading point.

- If the reason for rejection at the importing country is microbiological / residue contamination, samples for testing are collected batch / variety wise, following the sampling scale $(\sqrt{n}+1)/2=x$. Each individual sample is tested for the specific contaminant at two different laboratories. Preferably, one set of samples is tested at EIA lab. and the other set of samples at EIC approved laboratory. Testing is conducted in laboratories other than the lab that had conducted pre-export testing of the rejected cargo. Lab representative, who drew the sample(s) for pre export testing, is advised to be present. In addition to above, samples are drawn from the slabs / master carton (s) from which the samples are drawn by the importing country's health authority, if these are identifiable.
- In the case of rejection due to organoleptic factors, the returned consignment is inspected by EIA for all organoleptic factors. For this purpose, each variety of fishery product under a production batch(es) are treated as a lot and samples drawn as per the sampling scale covering all the grades. If the organoleptic evaluation of the returned consignment reveals any doubt as to the freshness of the product, the same is tested for TVB-N and TMA-N.
- In case of rejections which do not involve food safety / quality issues, the returned consignment is inspected and sampled, according to the nature of the complaint / rejection with a critical focus on the grounds of rejection. The objective of inspection / testing is to ascertain if the rejection is justified and if there was any lapse on the part of the approved processor / exporter.
- Actual testing charges and deputation fee are paid by the processor.
- If all the samples tested meet the specification, the concerned EIA in-charge take decision to release the consignment for export to countries other than EU & RF and other than the non-EU country where the consignment had

originally got rejected, provided it meets the requirements of the importing country.

- If any of the samples tested from the returned consignment shows unsatisfactory results, the processor either reprocess if it is possible to bring it to conformity without risk to health by scientifically accepted methods (For example, heat treatment, conversion of raw fishery products having microbiological issues into cooked products etc.) in presence of EIA officer or destroy the lots, if it cannot be brought into conformity without health risk by scientifically accepted method(s) (For example, issues related to residues)
- The method and schedule of reprocessing are furnished to the EIA for scrutiny. The processor offers the reprocessed consignment for inspection by EIA. The EIA inspect the reprocessed products batch wise for all microbiological parameters as well as organoleptic factors. If the reprocessed products are found export worthy on inspection, the lots are allowed for export.
- Inspection fee at the rate of 0.3% of the F.O.B. value of the consignment is charged for the reprocessed consignment for export. Testing fee is charged as per prescribed rates

9.1.14. Revocation of 'Internal Alert'

The in-charge of the EIA can revoke the 'Internal alert' imposed on the facility with an intimation to EIC under the following conditions.

1. The rejected consignment, if brought back to India tested for the contaminant / defect is found free of the contamination / defect, as evidenced by test reports or organoleptic inspection reports
2. The assessment report of the facility indicates satisfactory performance of the processing facility based on proper hygienic conditions and implementation of HACCP
3. The periodical monitoring conducted by EIA during the past 3 months indicates satisfactory performance and previous HACCP audit report is satisfactory;
4. Samples tested during the assessment visit pass
5. Five consecutive consignments / 3 days codes pass
6. Audit Report on the primary linkages, done by EIA(s), is satisfactory and suggestions are implemented.

9.2. Marine Products Exports Development Authority (MPEDA)

The MPEDA is an official Government authority constituted under the Marine Products Exports Development Authority Act 1972 (Section 13) functioning under the Department of Commerce, Government of India. It acts as a coordinating agency with State and Central Government establishments engaged in fishery production and allied activities. The registration of exporters, fishing vessels, and other processing entities is the main statutory function of MPEDA under Section 9(2) (b) and (h) of the MPEDA Act 1972. The registration is done for the following categories viz. Manufacturer Exporter, Merchant, Route through Merchant & Ornamental Fish Exporter and for entities such as Fishing Vessels, Processing Plants, Storage Premises, Conveyance, Pre-Processing centers, Live Fish Handling Centre, Chilled Fish Handling Centre, Dried Fish Handling Centre, Independent Cold Storages and ice plants

9.2.1. Role of MPEDA

- To promote the export of marine products.
- To develop and regulate off-shore and deep-sea fishing and to undertake measures for their conservation and management
- To register fishing vessels, processing plants, or storage premises for marine products and conveyances used for the transport of marine products
- To fix standards and specifications for marine products meant for export
- To render financial or other assistance (subsidy) to owners of fishing vessels engaged in off-shore and deep-sea fishing and to owners of processing plants or storage premises for marine products.
- To inspect marine products kept or handled in any fishing vessel, processing plant, storage premises, products to ensure the quality.
- To regulate and improve the marketing of marine products outside India
- To collect statistical data from persons engaged in the fishing, processors, exporters or such other persons engaged in fishing industry and to publishing the statistics.
- To training personnel in various aspects of the marine products industry

9.2.2. Additional Functions Undertaken by the MPEDA

The MPEDA also undertake the following functions.

- To assess the requirements of any machinery, equipment, and spares, including ancillary material, required for the handling and processing of marine products and, where necessary, recommend and arrange for import of such machinery, equipment, spares, and ancillary material
- To assess the standards of quality of indigenous processing equipment and recommend measures for their improvement

- To suggest the manufacture of new modern items of equipment required for the marine products industry
- To augment the availability of raw material for processing
- To assess the requirements of the marine products industry for cold-storage, transport, and other facilities and ensure the provision of such facilities
- To specify and enforce the layout of the processing plants equipment and other matters for maintaining the high quality of the marine products;
- To co-ordinate the demand and availability of reefer space and hold to regulate marine products' shipment from the existing and new ports.
- To undertake regulatory measures for conservation and management of fisheries on behalf of the Ministry of Agriculture of the Government of India.
- To undertake such other measures which shall directly or indirectly improve, organize, and develop the marine products industry with special reference to exports.

9.3. Certificate Systems for Fish and Fishery Products

9.3.1 Catch certificate-Illegal, unreported, and unregulated (IUU) fishing of EU

The MPEDA is an authorized agency for validating and issuing the catch certificates as per EU Regulation 1005/2008, for exporting fish and fishery products to the EU.

9.3.2. Health Certificate

The health certificate is issued for fishery products processed in establishments/ factory vessels approved and monitored by the EIA. The certificate is issued by EIC/EIAs to those products that confirms animal health standards and regulations meant for export. It states that the products are certified as suitable for human consumption. The validity of the health certificate is for a period of 90 days from the date of issue.

9.3.3. Seafood Import Monitoring Program – IUU fishing of the US

The US- Seafood Import Monitoring Program (SIMP) laydown conditions for imports of certain seafood products. The main aim of the program is to tackle the reporting requirements needed to prevent IUU caught or misrepresented seafood (seafood fraud) from entering the US. SIMP is a risk-based traceability program requiring the US importer to provide and report critical data of seafood from the point of harvest to entry into US commerce.

9.3.4. Certificate of Origin (CoO)

Certificate of Origin (CoO) is an autonomous commercial policy document used to promote and sustain mutual trade; and to develop economic co-operation among countries. A new online platform for issuance of preferential CoO is developed by the Directorate General of Foreign Trade of Ministry of Commerce. It is a single-point of access for CoO under all trade agreements. Three agencies are authorized to issue CoO, such as EIA (for all products), Directorate General of Foreign Trade (for all products), and MPEDA (for marine products).

9.3.5. DS -2031 Certificate to USA/Other Countries

DS 2031 certificate must accompany the shrimp product consignment exported to the USA as per Section 609 of Public Law 101-162 of USA. It certifies that the shrimp in the consignment is harvested in a manner not harmful to sea turtles or harvested by aquaculture. The MPEDA has online software to issue DS 2031 certificate to exporters since 6th April 2020 with digital signature and QR code for proof of authenticity.

9.3.6. International Commission for the Conservation of Atlantic Tunas (ICCAT) Certificate

International Commission for the Conservation of Atlantic Tunas is an inter-governmental fishery organization responsible for conserving tunas and tuna-like species in the Atlantic Ocean and its adjacent seas. The MPEDA is authorized to validate the ICCAT Swordfish Statistical document required to import fish like tuna, marlin, and swordfish to certain countries that border the Atlantic Ocean. The MPEDA has online validation of ICCAT certificate w.e.f. 01.04.2019.

9.3.7. Non-Radio-Activity Certificate

The MPEDA certifies fish and fishery products caught in Indian water for the presence of radioactivity either in traces or within acceptable limits, as per the importing country's requirement; and issues Non-Radio Activity Certificate to exporters.

9.3.8. Certificate of Legal Origin

National Fisheries and Aquaculture Service (Sernapesca) of Chile has authorized MPEDA to validate and issue the Certificate of Legal Origin by attesting that the goods in a particular export shipment is produced, manufactured, or processed, based on the request from exporters.

9.3.9. Duty-Free Import Certificate

The MPEDA is the Competent Authority to issue Duty Free Import Certificate to the exporters. The MPEDA certifies the export quantity and value of exporter's for a particular year, in order to get 1% duty benefit on Free on Board (FOB) value for importing the goods under the Customs Notification No.50/2017 dt. 30.06.2017.

9.3.10. Registration-cum-Membership Certificate (RCMC)

According to the Foreign Trade Policy, a Registration-cum-Membership Certificate (RCMC) is required for exporters to get benefit concerning customs and excise duty. The MPEDA is the Authorized Authority to issue RCMC for exporters of seafood products.

9.3.11. Certificate of Export

The certificate of export relating to the consignment is issued by the approved processor. The validity of the certificate is 15 days from the date of issue.

References

Act No. 22 of 1963. The Export (Quality Control and Inspection) Act, 1963 dated 24.08.1963, Government of India

Document No. EIC / F&FP / Ex.Inst. / March / 2012 / Issue 4 - Executive Instructions for Approval and Monitoring of Fish & Fishery Products for Export. Published by Export Inspection Council, Ministry of Commerce & Industry, Govt. of India

EIC website https://www.eicindia.gov.in/WebApp1/

MPEDA website https://mpeda.gov.in/

10

EU General Requirements for Import of Fishery Products

Introduction

The international fish trade increases day by day reaching to a tune of USD 151 billion in 2020 (FAO, 2022). Many developing countries like India gain benefits from this trade as income through net foreign exchange reached to a level of 8.09 billion USD in 2019-20. Seafood supplied by developing countries contribute to 77% of fish consumption worldwide. The European Union (EU) is the single largest market for fish and fishery products to a tune of 91.74 billion USD in 2022 as a consequence of an increased per capita consumption among its Member States. Spain, Italy, Germany and the United Kingdom are the largest importer of fish and fishery products.

To protect consumer health and to protect the territory from the introduction of animal and plant diseases, the European Community (EC) has laid down joint conditions for imports of foodstuffs. The Directorate General for Health and Consumer Protection (DG SANCO) is responsible for food safety in the EU. The import rules for fishery products are framed of high standards with respect to hygiene, consumer safety and animal health status. It is important that importing countries and food business operators (FBOs) must understand the fundamental principles and requirements of the European Food Law to ensure efficient and smooth imports. The EU system relies on Government to Government assurances without any intervention of any private type certification or standards. The Competent Authority (CA) in the country of origin is responsible for giving the official guarantees to the EU regarding compliance and conformity to all requirements of its legislation. The Export Inspection Agency (EIA) is the Competent Authority in India for issuing the required official guarantees for export of fish and fishery products. The EIA maintains a complete list of all establishments under its control.

The general requirements for export of fish and fishery products are given below:

10.1. Approved Establishments

All establishments in the capture or aquaculture production chain (hatcheries, farms, vessels, plants, cold-stores, etc) must be approved and registered with their national CA to export their products to the EU. Aquaculture products, including live bivalve mollusks must be produced within the approved zones or areas. The EIA is responsible for the **approval** of seafood processing establishments and the Marine Products Export Development Authority (MPEDA) of India is responsible for the **registration** of the seafood establishments.

The EIA sends the list of authorized establishments with the guarantee that they comply with the specific hygiene rules to the EC. The EC validates the list and communicates it to the Member States. The entire process takes **one to three months** for completion. The Directorate General for Health and Food Safety (DG-SANTE) provides the list of approved establishments. Four lists pertaining to fish and fishery products are available on the EU website.

- Section VII – Live bivalve molluscs – live, frozen or processed bivalve molluscs, echinoderms, tunicates, or marine gastropods (excluding the adductor muscle of Pectinidae). Live products must come from a harvest area that appears on the list.
- Section XIV – Gelatine – gelatin from fish
- Section XV – Collagen – collagen made from fish
- Section VIII – Fishery Products – all other fish and fishery products, including the adductor muscle of Pectinidae

10.1.1. Procedure for Getting EU Approval

To obtain EU approval, fish business operators (FBO) must submit the following to the CA in the country of origin i.e. EIA.

1. Registration or license number of the establishment
2. Full legal name as given in Certificate of Registration
3. Physical site address as given in Certificate of Registration
4. Species to be exported to the EU, indicating whether any aquacultured raw material is used
5. Name of the EU approval list

Fish processing operations meant for exports to the EU shall commence from the date request is submitted to DG-SANTE. However, the **Health Certificate** is issued by EIA, only after the validation and once the new or updated EU approved list enters into force.

10.1.2. Requirements for the Establishments

The EIA certifies compliance of the establishments to a series of requirements as listed in the public health attestation of the certificate. The first requirement is that the products must come from an establishment implementing a program based on the **HACCP principles in accordance with Regulation (EC) No. 852/2004**. The HACCP plan and its pre-requisite programs are the documents for the establishment of compliance systems.

The Regulation (EC) No. 853/2004 defines the key hygiene requirement of fish that are caught and handled on board vessels, landed, handled and where appropriate prepared, processed, frozen and thawed i.e. the whole seafood value chain must be under the control of EIA and must be put in compliance by FBOs.

10.1.2.1. Health Certificate

Each consignment of fishery products must be accompanied by a single, original, fully completed EU Health Certificate issued by the CA of the exporting nation (Regulation (EC) No. 854/2004). The EIA is responsible for the issue of Health Certificate for fish and fishery products in India. Council Directive 2006/88/EC of 24 Oct 2006 gives animal health requirements for aquaculture animals and products, thereof. The EU has set certification requirements for marketing of live bivalves, mollusks, echinoderms, tunicates, gastropods and fishery products to prevent introduction and spread of diseases within its territories. The list of susceptible species causing diseases is given below (Table 10.1).

Table 10.1. List of the susceptible fish, crustacean and molluscan species causing diseases

Disease	Susceptible species
Bonamia exitiosa	Oysters (Australian mud, Chilean flat)
Perkinsus marinus	Oysters (Pacific, Eastern)
Microcytos mackini	Oysters (Pacific, Eastern, Olympia flat, European flat)
Martelia refringens	Oysters (Australian mud, Chilean flat, European flat, Argentinian) and Mussel (Blue, Mediterranean)
Bonamic ostreae	Oysters (Australian mud, Chilean flat, Olympia flat, Asiatic, European flat, Argentinian)
Taura syndrome	Shrimps (Gulf white, Pacific blue, Pacific white)
Yellow head disease	Shrimps (Gulf brown, Gulf pink, Kuruma, Black tiger, Gulf white, Pacific blue, Pacific white)
White spot disease	All decapod crustaceans
Epizootic heamatopoietic necrosis (EHN)	Rainbow trout, Redfin perch

Koi herpes virus	Koi carp or Common carp
Infectious salmon anemia (ISA)	Rainbow trout, Atlantic salmon, Brown trout
Viral haemorrhagic septicaemia (VHS)	Herring, whitefish, pike, Pacific cod, Haddock, Pacific cod, Atlantic cod, Pacific salmon, Rainbow trout, Rockling, Brown trout, Turbot, Sprat, Grayling, Olive flounder
Infectious haematopoetic necrosis (IHN)	Chum salmon, Coho salmon, Masou salmon, Rainbow trout, Sockeye salmon, Pink salmon, Chinook salmon, Atlantic salmon

Animal health certificate is applicable, if the fishery products are of aquaculture origin. It is not applicable for fishery products not packaged for retail, live or whole un-eviscerated finfish or live crustaceans intended for further processing in the EU. It is not required for wild caught fish and products either slaughtered or eviscerated finfish and crustaceans packaged for direct consumption.

The **health certificate** must contain the description of the identity of the approved establishment, the type of fish, the quantity of product, and the final destination of the goods. The details of the product description must indicate whether the product originated from an aquaculture operation or is classified as a fishery product. The model health certificate is available in Regulations (EC) No. 1664/2006.

10.1.2.2. Catch Certificate

The EU established a system to prevent, deter and eliminate illegal, unreported and unregulated (IUU) fishing in the Council Regulation (EC) No. 1005/2008 of 29 Dec 2008. The countries exporting to the EU must provide a valid **catch certificate** attesting that fish and fishery products originate from non-IUU fisheries. It is applicable to all marine fishery products including live, fresh, chilled, frozen, prepared and preserved product forms. All freshwater and aquaculture products are exempted from this regulation. Some forms of mollusks such as scallops, mussels, oysters and snails are also excluded. The MPEDA is the responsible agency for issuing the catch certificate to the registered establishments in India.

10.1.2.3. Residue Monitoring Plan

The countries exporting aquaculture products to the EU need a specific approval stating that the establishment is in compliance with veterinary residue monitoring requirements, as outlined in Council Directive 96/23/EC. The third country (India) must submit a plan, setting out the guarantees on the monitoring of the groups of residues and substances referred in the Directive. There must be a centrally coordinated residue monitoring plan, which requires a description of the legislation governing the authorization, distribution and use of veterinary medicinal products, and also states the number of samples taken in accordance with the sampling levels and frequencies. The initial plan that has to be submitted by a third country must include:

1. Information on the structure of the CA
2. A description of the legislative framework
3. A list of approved laboratories for residue controls and the accreditation status of these laboratories
4. Rules covering the collection of official samples
5. Details on measures to be taken in the event of an infringement

The CA of the country of origin (EIA) submits the residue monitoring program to the EC for initial approval and it is evaluated annually for renewal. Aquaculture products consignment must have an approved residue control plan as listed in Commission Decision 2004/432/EC.

10.1.2.4. Import Controls

Fishery products and live animals imported into the territory of the EU are presented at the EU approved border inspection post (BIP) to undergo mandatory veterinary checks under the authority of an official veterinarian. There are around 300 authorized BIPs at ports, airports and land borders. A BIP must have adequate staff, facilities, storage premises, cold stores, and testing laboratory. The EC gives approval for BIP after inspection and assessment by the Food and Veterinary Office (FVO). The main EU legislation on border import controls concerning public and animal health are available in Council Directives 91/446/EEC and 97/78/EC, and Regulations (EC) No. 882/2004 and 854/2004.

10.1.2.4.1. Procedures at the BIPs

Inspection of consignments originating from third countries is carried out at BIP on all consignments at the first point of entry into the EU territory. Each consignment is notified to the relevant BIP with the information requested in the Common Veterinary Entry Document (CVED) using the computerized TRACES system, before its physical arrival (Directive 97/78/EC and Regulation (EC) No. 136/2004). The consignments are subjected to three types of checks.

1. **Documentary check** – it is done systematically and involves checking the export certificate accompanying the consignment
2. **Identity check** – it is done systematically and involves checking that the data on the export certificate is consistent with the product imported and

also verifying for the presence of required sanitary marks – country of origin, approval number, etc.

3. **Physical check** – it is done as appropriate and involves examination of the product, its packaging, information on the label, and the storage conditions.

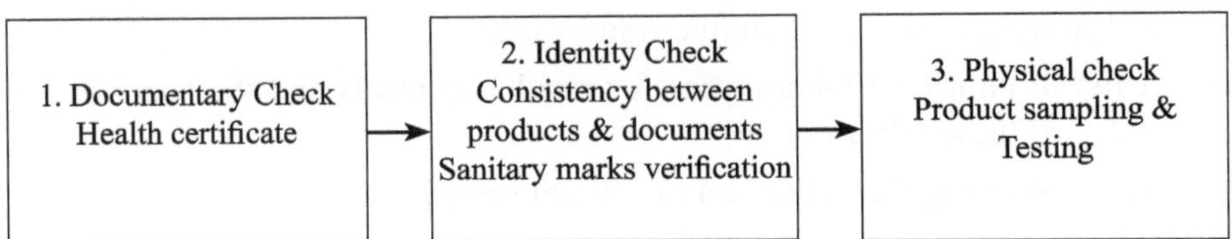

The documentary and identity checks are performed on all consignments. The physical check involves taking samples for laboratory testing on a random basis or on the basis of past records. Specific laboratory tests performed in accordance with the national monitoring plans to verify the product for residues, contaminants, pathogens or other substances of animal or public health concern. The frequency of physical check depends on the "**harmonization status**" of the third country of origin (Commission Decision 94/360/EC) i.e. it is determined for each category of product on the basis of the intrinsic risk and the results of checks carried out previously on the same product of the same origin. The physical check is done in approximately 20% of products such as those in hermetically sealed containers, fresh and frozen fish, and dry or/and salted. In case of other products and bivalve mollusks, the physical check is done in 50% of products. The CVED is then issued i.e. after the veterinary checks, to either release the consignment to be marketed freely in all member states, or for transit to third countries, or for storage in approved warehouses or to reject, if case of non-compliances with the EU law.

If the consignment is found to be in non-compliance with the EU legislation, the establishment is put on "**reinforced control status**", and the BIP notifies it to all member states and the EC through the internal notification system, known as the **Rapid Alert System for Food and Feed (RASFF)**. If an establishment is on reinforced control status, the next 10 consecutive shipments to any EU country will be automatically tested. The products are detained at BIP until the results are received. The seafood establishment in question is removed from the reinforced control list only after the next 10 shipments are without any positive results.

RASFF sends urgent notification, receives and responds in the shortest time possible. The EC publishes a weekly summary of the notifications. The notifications can be either "Alerts" or "Information". For instance, a total of 1175 alerts and 4118 notifications was given during 2019, and among which,

303 notifications are related to fish and fishery products. The majority of the risks notified is related to the presence of pathogenic microorganisms, mycotoxins, pesticide residues and heavy metals.

a. Alert notifications are given, when the food or feed presenting a risk is already on the market and immediate action is required.
b. Information notifications are sent, when a risk is identified but immediate action by other member states is not necessary, as the product has not reached their market, i.e. consignments stopped at the BIPs

When a notification pertains to an imported product, the CA of the country of origin (EIA) has to make a full investigation and reports back to the EU on their results and measures to avoid recurrences. The BIP detains the imported product, until the CA of the country of origin clarifies the situation. If a consignment is refused for non-compliance with the EU legislation, the exporter in the country of origin has three options:

1. Destroy the products
2. Re-dispatch the products to a non-EU country
3. Return the products to the originating country

10.2. Legal Standards for Monitoring Fishery Products Related to EU Requirements

Specific standards for fishery products are set out in the EU legislation. The legislation says that the international standards for fishery products given by Codex Alimentarius Commission (CAC) shall be adopted and applied by third countries, in case standards are not available in the EU. The consolidated list of Regulations providing requirements for various biological and chemical hazards associated with fishery products are given in Table 10.2.

Table 10.2. EU legislations related to requirements for various biological and chemical hazards

S.No.	Hazards	EU Regulations
1.	Microbiological requirements	Regulation EC No. 2073/2005
2.	Parasites	Regulation EC No. 1276/2011
3.	Food Additives	Regulation EC. No. 1333/2008
4.	Heavy metals, Dioxins	Regulation EC. No. 1881/2006
5.	Histamine	Regulation EC. No. 2073/2005
6.	Marine Biotoxins	Regulation EC. No. 853/2004; 2074/2005; 15/2011
7.	Polyaromatic hydrocarbons	Regulation EC. No. 1881/2006; 1327/2014
8.	Therapeutants & Antibiotics	Regulation EC. No. 470/2006; 37/2006

10.2.1. Prohibited Species

Certain fish species are naturally toxic and should not be marketed in the EU. Regulation (EC) No. 854/2004 set out the genera concerned with the families Tetradontidae, Molidae, Diodontidae and Canthigasteridae. Commission Regulation No. 1021/2008 extended the marketing restrictions for toxic fish of the family, Gempylidae, in particular *Ruvettus prestiosus* and *Lepidocybium flavobrunneum* (oil fish or escolar) that they must be placed on the market in wrapped/ packaged form and with appropriate information to the consumer on preparation/ cooking methods and on the risk related to the presence of substances with adverse gastrointestinal effects. The scientific names of the fishery products must appear on the label, along with the common names.

10.2.2. Organoleptic Quality Check

The Council Regulation (EC) No. 853/2004 under Section VIII – Fishery products, Chapter V gives the **health standards** for fishery products and specifies that FBO must carry out an organoleptic examination before placing the fish in market. The Council Regulation (EC) No 2406/96 of 26 November 1996 has laid down common marketing standards for certain fishery products, wherein products are classified as Extra (E), A or B for selachii, cephalopods and Norway lobsters; Extra (E) or A for shrimps; and Extra (E) in case of live Norway lobsters.

10.2.3 Total Volatile Basic

The Council Regulation (EC) No. 853/2004 states that unprocessed fishery products must not be placed on the market, if TVB-N or TMA-N exceeds the limits. The Council Regulation (EC) No 2074/2005 gives the limits for TVB-N specified for fishery products and the method of analysis. The TVB-N limits are set in the range of 25 to 35 mg of nitrogen /100 g of flesh for certain species of fish. The regulation has also set out a reference procedure for the method of TVB-N analysis.

10.2.4 Parasites of Public Health Concern

The Council Regulation (EC) No. 853/2004 has laid down specific hygiene rules on the hygiene of foodstuffs, and states that FBO must ensure that fishery products are subjected to a visual examination for the purpose of detecting visible parasites before placing them on the market. The Council Regulation (EC) No. 2074/2005 of 5 Dec 2005 under Chapter II gives the rules for visual inspection to detect parasites in fishery products.

Visible Inspection must be performed on a representative number of samples. Scale and frequency of inspection are determined by the type of product, geographic origin and their use. Inspection must be carried out by qualified persons on the abdominal cavity, livers, roes, fish fillets or slices. In case of manual evisceration, inspection must be done in a continuous manner by handler at the time of evisceration and washing. In case of mechanical evisceration, it is done by sampling i.e. 10 representative samples per batch. Visual inspection or candling are also followed.

10.2.5 Microbiological Criteria

The Council Regulation (EC) No 852/2004 of 29 April 2004 on the hygiene of foodstuffs specifies that it is necessary to establish microbiological criteria and temperature control requirements based on a scientific risk assessment and FBO shall adopt hygiene measures for compliance with microbiological criteria for foodstuffs. The Council Regulation (EC) No. 2073/2005 of 15 November 2005 has laid down the microbiological criteria for foodstuffs giving an obligation for FBO to ensure that foodstuffs follow the microbiological criteria through good hygiene practice and HACCP procedures, and for official controls to verify that the criteria laid down are met. Article 5(2) of the Regulation refers to ISO 18593 as a reference sampling method for sampling of processing areas and equipment. Point 3.1 of Chapter 3 of Annex I give the relevant standards of ISO and the guidelines of the Codex Alimentarius as reference methods.

Food safety criteria are available for shelled and shucked products of cooked crustacean and molluscan shellfish, cooked crustaceans and molluscan shellfish, crabmeat, live bivalve molluscs and ready-to-eat foods with target organisms, sampling plan, limits, analytical reference methods, stages where criterion applies, and actions to be taken, in case of unsatisfactory results.

10.3. Limit for Chemical Residues

The principal objective of Council Directive 96/23/EC is to detect illegal use of chemical substances in fish production as well as to detect the misuse of authorized veterinary medicinal products. The Directive lays down measures for their member states to monitor the substances and groups of chemical residues listed in Annex I. Annexure contains the substance groups classified under two main categories - Group A and Group B.

Group A lists the chemical substances having anabolic effect and unauthorized substances (i.e. most of the substances which are prohibited from use in food producing animals in the EU) and it is subdivided into 6 sub-groups (A1-A6). Group B contains a list of veterinary drugs and contaminants (i.e. residues of many pharmacologically active substances which may be authorized for use in food producing animals in the EU), which are listed in Annex I to III of Council Regulation (EEC) No. 2377/90. It also comprises organochlorine and organophosphate pesticides and other elements such as lead, cadmium and mercury. Annex II provide the list for each commodity (e.g. bovine animals, aquaculture animals, honey, milk, eggs etc.) under Group A and Group B subgroups.

10.3.1. Food Additives

The Council Regulation (EC). No. 1333/2008 of 16 Dec 2008 gives the permitted list of food additives (excluding coloring agents and sweeteners) with MRLs in ppm (Table 10.3). The major food additives covered under this list include benzoates, sorbates, sulphites, phosphates, Ca EDTA, boric acid, citrates and nitrates.

Table 10.3. The MRLs of food additives in certain fishery products.

Additives	Products	MRL (ppm)
Benzoic or Sorbic acid	Semi-preserved fish products (fish roe), Cooked shrimps	200
Benzoic or Sorbic acid	Salted, dried fish	200
Sulphites	Crustaceans – over 120 units - Penaeidae, Soleneridae, Aristedae	300
Sulphites	Dried, salted fish of Gadidae spp., Crustaceans – between 80 -120 units - Penaeidae, Soleneridae, Aristedae	200
Sulphites	Fresh and Frozen cephalopods, Crustaceans – upto 80 units Penaeidae, Soleneridae, Aristedae	150
Sulphites	Cooked crustaceans – Penaeidae, Soleneridae, Aristedae	50
Phosphates	Surimi	1000
Phosphates	Frozen fillets of unprocessed fish; Frozen molluscan shellfish (except scallops); Frozen crustaceans; Fish and crustacean pastes	5000
Calcium disodium EDTA	Canned and bottled fish; Frozen crustaceans and molluscs	75
Boric acid	Cavier (Sturgeon eggs)	4
Nitrates	Pickled herring and sprat	200
Sorbitol, Mannitol, Isomalt, Maltitol, Lactitol, Xylitol, Cirtates	Frozen unprocessed fish, crustaceans and molluscs	GMP*

*GMP – good manufacturing practices

10.3.2. Heavy Metals, Dioxins, and PCBs

Heavy metals such as lead, cadmium and mercury are present naturally in fish, while tin leaches from food can. Methyl mercury is in organic form and constitutes >90% of total mercury in fish. Predatory fish accumulate high amounts of heavy metals e.g. tuna, sturgeon, pike, pike perch, swordfish. The Council Regulation (EC). No. 1881/2006 of 19 Dec 2006 sets the maximum levels for lead, cadmium, mercury, tin, dioxins, and benzopyrene in fishery products (Table 10.4 a-c). The MRL for tin in all canned foods other than beverage is fixed as 200 mg/kg wet weight.

Table 10.4a. MRLs for cadmium in fishery products

Substrate	**MRL for cadmium (mg/ kg wet weight)**
Edible portion of all fishery products	0.05
Mackerel (*Scomber* sp), Tuna (*Thunnus* sp. *Katsuwonus pelamis*, *Euthynnus*), Bichique (*Sicyopterus lagocepalus*)	0.1
Muscle of bullet tuna (*Auxis* sp)	0.15
Muscle of Anchovy (*Engraulis* sp), Swordfish (*Xiphias gladius*), Sardine (*Sardina pilchardus*)	0.25
Crustaceans – muscle from appendages and abdomen; Crabs and crab-like crustaceans - muscle from appendages	0.5
Bivalve molluscs	1.0
Cephalopods (without viscera)	1.0

Table 10.4b. MRLs for mercury in fishery products

Substrate	**MRL for mercury (mg/ kg wet weight)**
Edible portion of all fishery products	0.5
Muscle meat of Angler fish, Atlantic catfish, Bonito, Eel, Emperor, Grenadier, Halibut, Kingklip, Marlin, Megrim, Mullet, Pink Cusk Eel, Pike, Plain Bonito, Poor Cord, Portuguese dogfish, Rays, Redfish, Sailfish, Scabbard fish, Sea bream, Shark, Snake mackerel, Sturgeon, Swordfish, Tuna	1.0

Table 10.4c. MRLs for lead in fishery products

Substrate	**MRL for lead (mg/kg wet weight)**
Edible portion of all fishery products	0.3
Crustacean, excluding brown meat of crab and excluding head and thorax meat of lobster and similar (Nephropidae and Palinuridae)	0.5
Bivalve molluscs	1.5
Cephalopods without viscera	0.3

The Commission Regulation (EC) No. 333/2007 of 28 March 2007 has laid down the methods of sampling and analysis for the official control of the levels of lead, cadmium, mercury, inorganic tin, 3-MPCD and benzopyrene in food stuffs. It provides the sampling method, sampling plan, sample preparation methods, and method of analysis, along with detection limits and limits of quantification (Table 10.5).

Table 10.5. The MRLs for dioxins permitted in fishery products

Fish products	Max level permitted – Sum of dioxins	Max level permitted – Sum of dioxins and dioxin like PCBs	Max level permitted – Sum of PCB congeners
Muscle meat of fish and seafood excluding the following: - Wild caught fish, - Wild caught freshwater fish, - Fish liver and derived products, - Marine oils	3.5 pg/g wet weight	6.5 pg/g wet weight	75 ng/g wet weight
Muscle meat of wild caught eels (*Anguilla anguilla*) and products thereof	3.5 pg/g wet weight	10.0 pg/g wet weight	300 ng/g wet weight
Muscle meat of wild caught freshwater fish, and products thereof	3.5 pg/g wet weight	6.5 pg/g wet weight	125 ng/g wet weight
Fish liver and products derived thereof	-	20 pg/g wet weight	200 ng/g wet weight
Fish oil intended for human consumption	1.75 pg/g fat	6.0 pg/g fat	200 ng/g fat

10.3.3. Marine Biotoxins

Some fish species contain biotoxins. Ciguatera fish poisoning (CFP) is a frequently encountered toxic syndrome due to the consumption of ciguatoxin containing fish. Ciguatoxin (CTX) are mainly produced by a dinoflagellate, *Gambierdiscus toxicus*, which is eaten by herbivorous fish first and then by larger carnivorous fish. The CTX accumulates in liver, muscles, skin and bones of fish. Fish associated with CFP in Indo-Pacific region are Great barracuda (*Sphyraena barracuda*) and Giant moray (*Gymnothorax javanicus*).

The Council Decision 2002/226/EC of 15 March 2002 mentions on the special health checks to be done for harvesting and processing of certain bivalve mollusks with a new level of amnesic shellfish poison (ASP). Accordingly, Chapter V – Health Standards for Fishery Products of Regulation (EC) No. 853/2004 has given standards stating that fishery products containing biotoxins such as ciguatoxin or muscle-paralysing toxins must not be placed on the market. The FBO must ensure that live bivalve molluscs meet the following standards i.e. they must not contain marine biotoxins in total quantities exceeding the following limits:

a. For Paralytic Shellfish Poison (PSP), 800 µg of STX/kg
b. For Amnesic Shellfish Poison (ASP), 20 mg of domoic acid /kg;
c. For okadaic acid, dinophysis *toxins* and pectenotoxins together, 160 µg of okadaic acid equivalents/kg;
d. For yessotoxins, 1 mg of yessotoxin equivalent/kg
e. For azaspiracids, 160 µg of azaspiracid equivalents /kg

The Commission Regulation (EC) No. 15/2011 of 10 January 2011 amended the Regulation (EC) No 2074/2005 with regard to the recognized testing methods for detecting marine biotoxins in live bivalve mollusks. It states EU-RL LC-MS/MS method shall be the reference method for the detection of marine toxins as referred to in Regulation (EC) No 853/2004 and the method shall determine at least the following compounds:

- Okadaic acid group toxins: OA, DTX1, DTX2, DTX3 including their esters,
- Pectenotoxins group toxins: PTX1 and PTX2,
- Yessotoxins group toxins: YTX, 45 OH YTX, homo YTX, and 45 OH homo YTX,
- Azaspiracids group toxins: AZA1, AZA2 and AZA3.

The Commission Regulation (EU) No. 786/2013 of 16 August 2013 amended Annex III of the Regulation (EC) No. 853/2004 with regard to the permitted limits of yessotoxins in live bivalve mollusks. The latest Commission Regulation (EU) 2017/1980 of 31 October 2012 amended Annex III of the Regulation (EC) No. 2074/2004 with regard to paralytic shellfish poison (PSP) detection method.

10.3.4. Polyaromatic Hydrocarbons

Smoking is one of the methods used for preserving fish. Many antibacterial compounds are derived through partial combustion of wood. Some of them are carcinogenic. Smoke flavours contain benzopyrene and sum of benzo (a) pyrene, benzo (a) anthracene, benzo (b) fluoranthene and chrysene, which are markers to know the effect of carcinogenic polyaromatic hydrocarbons (PAH). The EU has established limits for the amount of PAH in different fishery products in Regulation (EC) No. 1881/2006 by setting maximum levels for certain contaminants in foodstuffs (Table 10.6). Sampling method, sample treatment, preparation, and performance criteria for analytical method for testing benzo(a) pyrene are given in Regulation (EC) No. 333/2007.

Table 10.6. The maximum permitted level of PAHs in different fishery products

Fish products	Maximum permitted level Benzo (a) pyrene (µg/kg)	Maximum permitted level Sum of benzo (a) pyrene, benzo (a) anthracene, benzo (b) fluoranthene and chrysene (µg/kg)
Oils and fats intended for direct human consumption or use as an ingredient in food	2.0	10.0
Muscle meat of smoked fish and smoked fishery products, excluding products listed below	2.0	12.0
Smoked sprats and canned smoked sprats (*Sprattus sprattus*)	5.0	30.0
Bivalve mollusks (fresh, chilled or frozen)	5.0	30.0
Bivalve mollusks (smoked)	6.0	35.0

The maximum level for smoked crustaceans applied to muscle meat from appendages and abdomen. In case of smoked crabs and crab-like crustaceans, it applies to muscle meat from appendages. The companies willing to export fish products to the EU are responsible for monitoring of PAHs in their products to meet the criteria.

10.3.5. Antibiotics and Therapeutants

The EU requirements specifying permitted and not permitted veterinary medicines and conditions for their use in food animals are revised and replaced with the Regulation (EC) No. 37/2010 giving the list of pharmaceutically active substances and the maximum residue limits in foodstuffs of animal origin (Table 10.7).

Table 10.7. The MRLs of list of substances allowed for application in fish in aquaculture production

S.No	Compound	MRL (µg/kg)
1.	Amoxicillin	50
2.	Ampicillin	50
3.	Chlortetracycline	100
4.	Cypermethrin (Salomindae)	50
5.	Diflubenzuron (Salmonidae)	1000
6.	Flumequin (finfish)	600
7.	Flumequin (trout)	500
8.	Flumequin (all species)	200

S.No	Compound	MRL (µg/kg)
9.	Oxolinic acid	100
10.	Oxy tetracyline	100
11.	Sarafloxacin	30
12.	Sulfonamides	100
13.	Teflubenzuron (Salmonidae)	500
14.	Tetracyline (fish)	200
15.	Trimethroprim	50

There are pharmacologically active substances for which no maximum limits are fixed. Table II in EC Regulation No. 37/2010 gives the list of prohibited substances that should not be applied to food animals. The compounds are *Aritolochia* spp and the preparations thereof, chloramphenicol and derivatives (thiamphenicol), chloroform, chlorpromazine, colchicine, dapsone, dimetridazole, metronidazole, nitrofurans (including furazolidone) and ronidazole.

There are several other substances not permitted for use in production of food animals in the EU, which are not currently listed in Table II. They are malachite green, crystal violet, growth promoting hormones, etc.

In order to harmonize the analytical methods used by laboratories for residue testing, the EC has established minimum required performance limits (MRLs). Commission Decision 2002/657/EC has given guidelines concerning the performance of analytical methods and the interpretation of results. For instance, the methods used for the detection of malachite green and leuco malachite green in aquaculture products are required to detect these compounds at a MRPL of 2∝g/kg.

10.3.5.1. Interpretation of Test Reports

The Commission Decision 2002/657/EC provides rules for the analytical methods to be used in the testing of official samples and specifies common criteria for the interpretation of analytical results of official control laboratories. The substances concerned are the following:

1. Mycotoxins in foodstuffs (Commission Regulation (EC) No 401/2006).
2. Dioxins and dioxin-like PCBs in foodstuffs (Commission Regulation (EC) No 1883/2006)
3. Lead, Cadmium, Mercury, Inorganic tin, 3-MCPD and benzopyrene in foodstuffs (Commission Regulation (EC) No 333/2007)

Article 3 of the Decision states the official samples are analyzed by using the methods that are documented in test instructions according to ISO 78-2 (6), validated according to the procedures, and comply with the relevant MRPLs. The Member States must ensure the results of the analysis by monitoring tests and calibration results according to ISO 17025. Article 6 gives the interpretation of the results, as given below.

1. The results of an analysis must be considered non-compliant, if the decision limit of the confirmatory method for the analyte is exceeded.
2. If a permitted limit is established for a substance, the decision limit is the concentration above which it can be decided with a statistical certainty of 1-α that the permitted limit has been truly exceeded.
3. If no permitted limit is established for a substance, the decision limit is the lowest concentration level at which a method can discriminate with a statistical certainty of 1-α that the particular analyte is present
4. For substances listed in Group A of Annex 1 of Directive 96/23/EC, the "α" error shall be 1% or lower and for all other substances, the "α" shall be 5% or lower.

References

Commission Implementing Regulation (EU) 2019/628 of 8 April 2019 concerning model official certificates for certain animals and goods and amending Regulation (EC) No 2074/2005 and Implementing Regulation (EU) 2016/759 as regards these model certificates

Commission Regulation (EC) No 1664/2006 of 6 November 2006 amending Regulation (EC) No 2074/2005 as regards implementing measures for certain products of animal origin intended for human consumption and repealing certain implementing measures

Commission Regulation (EC) No 1881/2006 of 19 December 2006 setting maximum levels for certain contaminants in foodstuffs

Commission Regulation (EC) No 2073/2005 of 15 November 2005 on microbiological criteria for foodstuffs

Commission Regulation (EU) No 37/2010 of 22 December 2009 on pharmacologically active substances and their classification regarding maximum residue limits in foodstuffs of animal origin

Council Directive 2006/88/EC of 24 October 2006 on animal health requirements for aquaculture animals and products thereof, and on the prevention and control of certain diseases in aquatic animals

Council Regulation (EC) No 1005/2008 of 29 September 2008 establishing a Community system to prevent, deter and eliminate illegal, unreported and unregulated fishing.

Regulation (EC) No 1333/2008 of the European Parliament and of the Council of 16 December 2008 on food additives

Regulation (EC) No 178/2002 of the European Parliament and of the Council of 28 January 2002 laying down the general principles and requirements of food law, establishing the European Food Safety Authority and laying down procedures in matters of food safety

Regulation (EC) No 852/2004 of the European Parliament and of the Council of 29 April 2004 on the hygiene of foodstuffs

Regulation (EC) No 853/2004 of the European Parliament and of the Council of 29 April 2004 laying down specific hygiene rules for food of animal origin

Regulation (EC) No 854/2004 of the European Parliament and of the Council of 29 April 2004 laying down specific rules for the organisation of official controls on products of animal origin intended for human consumption

11

Nutrition Facts and Nutrition Labelling Legislations

Introduction

Nutrition is the study of how food and nutrients affect the body's health and well-being. It encompasses all aspects of food, from the food production, processing, consumption and digestion by the body. A balanced diet included all the essential nutrients such as carbohydrates, protein, fat, vitamins and minerals for maintaining good health. The recommended daily intake of nutrients varies depending on the age, gender and physical activity of a person. The nutritional need also varies depending on the lifestyle and medical condition of a person. The nutrition facts are based on recommended daily intakes for various nutrients, which vary based on factors such as age, sex, and physical activity. These facts are given on food packaging or label, providing important information on the serving size, calories, macronutrients and micronutrients, and other ingredients.

Nutrition labels are mainly intended to guide the consumer in food selection. Nutrition information provided on labels must be true and should not mislead consumers. Nutrition labelling regulations are provided by countries to help manufacturers. The Codex guidelines on Nutrition Labelling provide guidance to member countries to develop or update their national regulations and to encourage harmonization of national standards with international standards. According to it, no food should be described or presented in false, misleading or deceptive way. Nutrition labelling cannot solve nutrition problems, instead it is one of the elements of nutrition policy.

11.1. Nutrition Facts

Nutrition Facts means the nutritional information required for most of the packaged food in many countries giving the energy value and nutrients contained in the food. Nutrition facts labels are required by regulation similar to other types of food labels. It provides detailed information on the nutrient content of the food such as the amount of fat, sugar, sodium and dietary fiber.

11.1.1. Overview of the Nutrition Label

A sample label for cooked shrimps is given in Fig 1. The information in the top section of the label varies with each food. It contains product-specific information such as serving size, calories, and nutrient information. The bottom section contains a footnote that explains the % Daily Value and it gives the number of calories used for general nutrition.

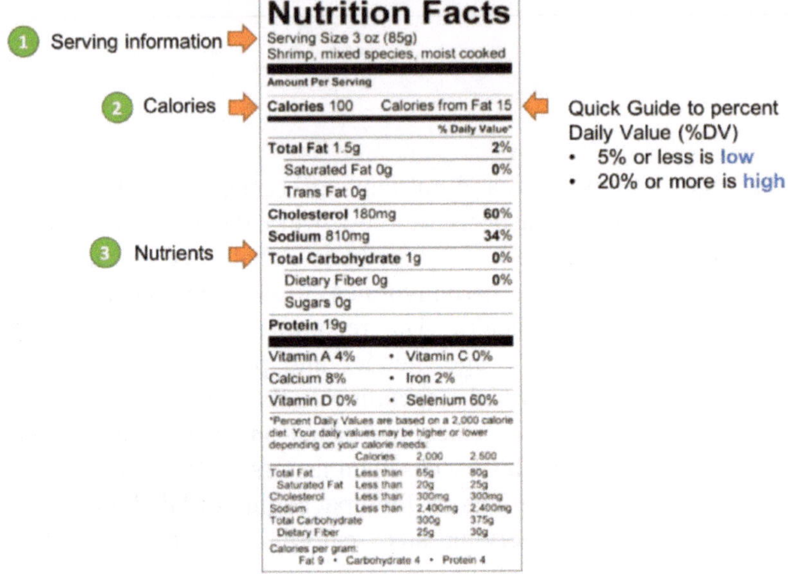

11.1.1.1. Serving Information

In the Nutrition Facts label, the number of servings in the package and the serving size are the serving information. Serving sizes are provided in units such as cups or pieces or ounces, followed by the metric amount, e.g. grams (g). The serving size reflects the amount that a person normally eats or drinks. The number of calories refers to the serving size. In the given sample label, one serving of cooked shrimps equals to 3 ounces.

11.1.1.2. Calories

Calories are measure of energy derived from a serving of a food. In the given sample label, there are 100 calories in one serving of cooked shrimps. This package has one serving size, while several other packages have 3 or 4 servings. The total number of calories is balanced to maintain a healthy body weight. The general guide for nutrition advice is to have 2000 calories a day. The calorie needs vary depending on our age, sex, height, and physical activity level.

11.1.1.3. Nutrients

In the given sample label, the key nutrients that impact our health are given. A person can decide on food that contains more or less of the nutrients.

Bad nutrients

Saturated fat, sodium and added sugars are nutrients having adverse health effects and are identified as nutrients to be consumed less. High consumption of saturated fat and sodium is associated with an increased risk of cardiovascular disease and high blood pressure. More consumption of added sugars makes us hard to meet other important nutrient needs within the calorie limits. Total sugars include sugars naturally present in many foods. No Daily Reference Value (DRV) is established for total sugars. Added sugars include sugars that are added during the processing of foods such as sucrose or dextrose, sweeteners, sugars in syrups and honey, and sugars in concentrated fruit or vegetable juices. Added sugars are given on the label as the number of grams of total sugars. In the given sample label, sugar is 0 g, so there is no added sugar.

Good Nutrients

Dietary fiber, vitamin D, calcium, iron and potassium are good nutrients on the label. Consumption of a diet high in dietary fiber increases the frequency of bowel movements, lowers blood glucose and cholesterol levels, and reduces calorie intake. Diet higher in vitamin D, calcium, iron, and potassium reduces the risk of developing osteoporosis, anemia, and high blood pressure. In the given sample label, dietary fiber is 0g, as seafood is not good sources of fiber. Other good nutrients available are vitamin A, vitamin D, iron and selenium.

11.1.1.4. The Percent Daily Value

The Percent Daily Value (%DV) is the percentage of the daily value for each nutrient in a serving of the food. The DV are reference amounts of nutrients, usually expressed in grams, milligrams or micrograms, to be consumed or not to be exceeded each day. The % DV shows the amount of nutrient in a serving of a food that contributes to a total daily diet. The % DV helps to determine, if a serving food is high or low in a nutrient. Some nutrients like total sugars, and trans-fat, do not have a %DV.

General Guide to % DV

5% DV or less of a nutrient per serving is considered **Low**

20% DV or more of a nutrient per serving is considered **High**

So, foods with higher %DV for dietary fiber, vitamin D, calcium, iron, and potassium; but lower %DV for saturated fat, sodium and added sugar must be chosen. The DV based on a 2000 calorie diet is given below:

Nutrient	Daily Value
Saturated Fat	Less than 20 g
Sodium	Less than 2300 mg
Dietary Fiber	At least 28 g
Added sugars	Less than 50 g
Vitamin D	At least 20 µg
Calcium	At least 1300 mg
Iron	At least 18 mg
Potassium	At least 4700 mg

"Upper limit" indicates that it is recommended to eat less than the DV nutrient amounts listed per day. For example, the DV for saturated fat is 20 g. "Lower" limit indicates that it is recommended to eat at least the DV of nutrient listed per day. For example, the DV for dietary fiber is 28 g.

Trans fat and Total Sugars are not listed on %DV on the Nutrition Facts Label. Protein is only listed on %DV in specific situation. Diets higher in trans-fat are associated with increased blood levels or low-density lipoprotein cholesterol, which are associated with an increased risk of cardiovascular disease. Scientific evidence also indicates that protein intake is not a public health concern for adults and children over 4 years of age.

11.1.2. Types of Nutrition Facts Label

Many Nutrition Facts Labels are available in the market. The two alternate formats permitted for use are dual column label and single ingredient sugar label.

11.1.2.1. Dual Column Label

For certain products that are larger than a single serving but that could be consumed in one sitting or multiple sittings, manufacturers provide "dual column" labels to indicate the amounts of calories and nutrients on both "per serving" and "per package" or "per unit" basis. It allows consumers to easily identify the calories and nutrients need to be consumed from the package/ unit at one time. The given sample label below shows the calories and other nutrients in one serving and in one package (3 servings).

Nutrition Facts

3 servings per container

Serving size **3 pretzels (28g)**

	Per serving		Per container	
Calories	**110**		**330**	
		% DV*		% DV*
Total Fat	0.5g	**1%**	1.5g	**3%**
Saturated Fat	0g	**0%**	0g	**0%**
Trans Fat	0g		0g	
Cholesterol	0mg	**0%**	0mg	**0%**
Sodium	400mg	**17%**	1200mg	**52%**
Total Carb.	23g	**8%**	69g	**24%**
Dietary Fiber	2g	**7%**	6g	**21%**
Total Sugars	<1g		3g	
Incl. Added Sugars	0g	**0%**	0g	**0%**
Protein	3g		9g	
Vitamin D	0mcg	0%	0mcg	0%
Calcium	10mg	0%	30mg	2%
Iron	1.2mg	6%	3.6mg	18%
Potassium	90mg	0%	270mg	5%

* The % Daily Value (DV) tells you how much a nutrient in a serving of food contributes to a daily diet. 2,000 calories a day is used for general nutrition advice.

11.1.2.2. Single Ingredient Sugar Labels

Package of products such as pure honey, pure maple syrup or packages of pure sugar are not required to include a declaration of the number of grams of added sugars in a serving of the product, but still include a declaration of the %DV for Added Sugars. So, manufacturers are encouraged to use the "†" symbol immediately following Added Sugars per Daily Value. It explains the amount of added sugars that one serving of the product contributes to the diet and the contribution of a serving toward the %DV for Added Sugars. The given sample label is of a single ingredient sugar e.g. honey.

Nutrition Facts

16 servings per container

Serving size 1 Tbsp. (21g)

Amount per serving

Calories	**60**
	% Daily Value*
Total Fat 0g	**0%**
Saturated Fat 0g	**0%**
Trans Fat 0g	
Cholesterol 0mg	**0%**
Sodium 0mg	**0%**
Total Carbohydrate 17g	**6%**
Dietary Fiber 0g	**0%**
Total Sugars 17g	
	34%†
Protein 0g	
Vitamin D 0mcg	0%
Calcium 0mg	0%
Iron 0mg	0%
Potassium 0mg	0%

* The % Daily Value (DV) tells you how much a nutrient in a serving of food contributes to a daily diet. 2,000 calories a day is used for general nutrition advice.

† One serving adds 17g of sugar to your diet and represents 34% of the Daily Value for Added Sugars.

11.2. Nutrition Labelling

Nutrition labelling is a description intended to inform the consumer of the nutritional properties of a food. It was developed in 1970s due to increased concern about health and nutrition.

11.2.1. Principles for Nutrition Labelling

A. Nutrient Declaration

A suitable profile of nutrients with nutritional importance must be provided to consumers. The information must convey the quantity of nutrients contained in the product.

B. Supplementary Nutrition Information

The content of supplementary nutrition information varies from one country to another and within any country from one target population group to another according to the policy and the needs of the target groups.

C. Nutrition Labelling

Nutrition labelling should not deliberately imply that a food that carries labelling has many nutritional advantages over a food which is not so labelled.

11.2.2. National Legislation on Labelling

In India, the law does not permit any wholesale dealer, distributor or importer to sell, distribute or display any commodity in the packaged form that is misbranded and non-compliant with the statutory labelling requirements. The **labelling requirements** for packaged commodities in India are governed under the Legal Metrology Act, 2009 and the Legal Metrology (Packaged Commodities) Rules, 2011. The **labelling of food products** is governed under the Food Safety and Standards Act, 2006 and the Food Safety and Standards (Packaging and Labelling) Regulations, 2011. The labelling requirements in India vary in different states. Certain states require to provide the statutory warning in the local language. There are different labelling requirements for the retail and the wholesale packages. Non-compliance of the statutory prescribed requirements and the regulatory registrations / licenses leads to monetary penalty.

The FSSAI Regulation Chapter 2 Packaging and Labelling gives the containers used for packaging including utensils, plastic materials and their specifications. Packaging requirements are given separately for each product– milk, edible oils, fruits, vegetables, meat, and drinking water. Section 3 gives the labelling requirements. It states every pre-packaged food should carry a label containing information in English or Hindi or any other language that is not false, misleading or deceptive. The content on the label must be clear, prominent, indelible and legible.

Labelling requirements of pre-packaged foods

The labelling requirements of every package of food must carry the name of the food, list of ingredients in descending order of their composition by weight or volume, and with specific name. Where the fish constitutes an ingredient of another food, it refers to all species of fish. The labelling and presentation of such food does not refer to a species of fish. If a product contains two or more ingredients, a compound ingredient shall be declared. Added water should be declared especially in case of dehydrated or condensed food.

Nutritional information

Nutritional Information or nutritional facts per 100 gm or 100ml or per serving of the product should be given on the label. It must contain (i) energy value in kcal; and (ii) the amounts of protein, carbohydrate (specify quantity of sugar) and fat in gram (g) or ml; and (iii) the amount of any other nutrient for which a nutrition or health claim is made. If a claim is on the fatty acids or cholesterol, the amount of saturated fatty acids, monounsaturated fatty acids and polyunsaturated fatty acids in gram (g) and cholesterol in milligram (mg) should be declared, and the amount of trans fatty acid in gram (g) also be declared. Numerical information on vitamins and minerals is declared in metric units The amount in gram (g) or milliliter (ml) per serving must be declared, besides the serving measure. If a claim for food to be enriched with nutrients, the quantities of, minerals, proteins, vitamins, metals or their compounds, amino acids or enzymes added shall be given on the label.

The compliance to quantity of declared nutrients on the label is according to the established practices. The food, in which hydrogenated vegetable fats or bakery shortening is used must declare on the label their use. A health claim of 'trans-fat free' means where the trans-fat is less than 0.2 gm per serving of food. A claim of 'saturated fat free' means where the saturated fat does not exceed 0.1 gm per 100 gm or 100 ml of food.

11.2.3. International Legislation on Labeling

11.2.3.1. Codex Alimentarius Commission

Codex Alimentarius CAC/GL 2-1985 (Rev 1-1993) gives the International guidelines on nutrition labelling. The purposes of these guidelines are to provide consumers with information to make their own food choices, to encourage improved formulation of foods, and to prevent deceptive nutrition labeling (FAO/WHO, 1987).

Nutrient declaration

Codex guidelines contain advisory criteria and standards for the nutrition labeling of foods. The guidelines state that nutrient declarations should be mandatory on any food for which a nutrition claim is made (Codex sections 3.1–5.1), but voluntary for all other foods. A nutrition claim must be declared for the energy value, the amounts of protein, available carbohydrate (not fiber), fat, and any other nutrient. A carbohydrate claim triggers the disclosure of the amount of total sugars. The amounts of starch and other carbohydrate constituents may be listed. A dietary fiber claim must disclose the amount of dietary fibre. A fatty acid claim must declare saturated and polyunsaturated fatty acids. Vitamins and minerals of nutritional importance to individual and present in significant amounts, defined as 5 % of the recommended intake, should be declared.

Calculation of nutrients

The amount of energy is calculated by using the following conversion factors:

Carbohydrates	4 kcal/g - 17 kJ
Protein	4 kcal/g - 17 kJ
Fat	9 kcal/g - 37 kJ
Alcohol (Ethanol)	7 kcal/g - 29 kJ
Organic acid	3 kcal/g - 13 kJ

The amount of protein is calculated from total Kjeldahl nitrogen x 6.25.

Presentation of nutrient content

- The declaration of nutrient content should be numerical.
- Information on energy value should be expressed in kJ and kcal per 100 g or per 100 ml or per package.
- Information on the amounts of protein, carbohydrate and fat should be expressed in g per 100 g or per 100 ml or per package. Information on protein may be expressed as percentages of the Nutrient Reference Value.
- Information on vitamins and minerals should be expressed in metric units and/or as a percentage of the Nutrient Reference Value per 100 g or per 100 ml or per package.
- In addition, this information may also be given per serving as quantified on the label
- Carbohydrate claim should declare the total carbohydrate content as follows: "Carbohydrate... g, of which sugars "x".. g"; where "x" represents the specific carbohydrate constituent.
- Fatty acid claim should declare the total fat content as follows: "Fat….g, of which polyunsaturated …g and saturated ….g"

Nutrient Reference Values

Nutrient Reference Values should be used for labelling purposes in the interests of international standardization and harmonization:

There are two types of NRVs:

1. Nutrient Reference Values-Requirments (NRVs-R)
2. Nutrient Reference Value - Non-communicable Disease (NRVs-NCD)

1. NRVs-R

Protein	50 g
Vitamins	
Vitamin A	800 μg RAE or RE
Vitamin D	5-15 μg
Vitamin C	100 mg
Vitamin K	60 μγ
Vitamin E	9 mg
Thiamin	1.2 mg
Riboflavin	1.2 mg
Niacin	15 mg NE
Vitamin B_6	1.3 mg
Folic acid	400 mg DFE
Vitamin B_{12}	2.4 μg

Panthothenate	5 mg
Biotin	30 μg
Minerals	
Calcium	1000 mg
Magnesium	300 mg
Iron	14 mg (15% dietary absorption, diets rich in meat, fish, fruits, etc.) 22 mg (10% dietary absorption, diets rich in cereals, roots or tubers, some fish)
Zinc	11 mg (30% dietary absorption, mixed diet, lacto-ovo vegetarian diets >90% flour) 14 mg (12% dietary absorption, cereal based diets)
Iodine	150 μg
Copper	900 μγ
Selenium	60 μg

2. NRVs-NCD

Intake levels not to exceed	
Saturated fatty acids	20 g
Sodium	2 000 mg
Intake levels to achieve	
Potassium	3 500 mg

Tolerances and Compliance

Tolerance limits is set in relation to public health concerns, shelf-life, accuracy of analysis, processing variability and inherent lability and variability of the nutrient in the product, and, according to whether the nutrient is added or is naturally occurring in the product.

Criteria for Legibility of Nutrition Labelling

Nutrition labelling whether applied on a mandatory or voluntary basis, the principles of the General Standard for the labelling of Prepackaged Foods (CXS-1,-1985, CAC, 2018) should be applied. Specific features are intended to enhance the legibility.

Supplementary Nutrient Information

The Codex guidelines include provisions for supplementary information to be given in addition to the nutrient declaration to increase the consumer's understanding of the nutritional value of their food and to assist in interpreting the nutrient declaration. The use of supplementary nutrition information on food labels is optional. Food group symbols or other pictorial or colour presentations can be used without the nutrient declaration. Consumer education programs are considered essential to enhance nutrition labeling.

11.2.3.2. United States of America

In the USA, nutrition labelling became a voluntary nutrition labeling scheme in 1975. Only products making nutritional claims or with added nutrients were required to have a nutrition label. Later, the United States Nutrition Labeling and Education Act (NLEA) was enacted in 1990 based on the recommendations of USFDA. The NLEA requires that packaged foods have a Nutrition Facts Panel, listing selected nutrients per serving in grams as well as per Recommended Daily Allowance (RDA). Later, similar rules were adopted for meat, poultry, and eggs by the US Dept of Agriculture in 1995. The FDA does not require any specific format to be used in the Nutrition Facts Label, but it mandates only a single easy-to-read type. It is mandatory that information listed to be written in English, including name of the product, net quantity, serving size and number of servings per package, nutrition facts, ingredient list, and name of the manufacturer or distributor.

The Nutrition Facts panel was introduced in 1993. The label has a standard serving sizes, calories, and the list of nutrients, including % daily value. The nutrients include, total fat, sodium, carbohydrates, and protein. A total of 15 nutrients are given calories, calories from fat, fat, saturated fat, trans fat, cholesterol, sodium, carbohydrates, dietary fiber, sugars, protein, vitamin A, vitamin C, calcium and iron. Trans fat is required to be listed under saturated fat from January 2006.

A new design of Nutrition Facts label was made by the FDA based on current scientific information, nutrition research and input from the public, to help the people to make their own food choices on May 20, 2016. For food and dietary supplement labeling, the amounts of vitamins and nutritionally essential minerals in a serving are expressed as a percent of Daily Value (%DV). The Reference Daily Intake (RDI) values, are derived based on the Recommended Dietary Allowances (RDA) for each nutrient. The RDA for vitamins C, D, E, K, calcium, phosphorus, magnesium and manganese are 50% higher than the older Daily Values used in labeling. A table of the old and new adult Daily Values is provided at RDI.

In addition to the nutrition label, products can display certain nutrition information or health claims on packaging. Health claims are only allowed by the FDA for "eight diet and health relationships based on proven scientific evidence", which include 1) calcium and osteoporosis, 2) fiber-containing grain products, 3) fruits and vegetables and cancer, 4) fruits, vegetables, and grain products that contain fiber—particularly soluble fiber—and the risk of coronary heart disease, 5) fat and cancer, 6) saturated fat and cholesterol and coronary heart disease, 7) sodium and hypertension, and 8) folate and neural tube defects.

11.2.3.3. European Union

In European Union, nutritional labelling is regulated by the Commission Directive 2008/100/EC for food. The Directive recommends daily allowance, energy conversion factors and definition. The new Regulation came in force in 2011 under EC Regulation 1169/2011 and has been mandatory for pre-packaged foods since December 2016. The information is labelled as "Nutrition Information". The panel is optional and if given, the prescribed format must be followed. The values are given as 100 g (3oz) or 100 ml (3.5 oz) of a product, also defined as "serving". Energy values are given in kcal or kJ. The mandatory information is energy, fat, saturates, carbohydrates, sugars, protein and salt. Optional list includes mono-unsaturates, polyunsaturates, polyols, starch, fibre, vitamins and minerals.

Health claims and Nutrition claims are harmonized in the EU through Regulation 1924/2006 amended as Regulation No. 1047/2012 and Regulation No. 1048/2012. Health claims are allowed on getting approval from European Food Safety Authority.

11.2.3.4. Australia

Australia and New Zealand use a nutritional information panel containing energy, protein, total fat, saturated fat, carbohydrates, sugars, and sodium along with serving per package and serving size.

11.2.3.5. Canada

Nutrition Label was introduced in Canada as part of regulations in 2003 and become mandatory in pre-packaged food products in 2005. The nutrition label as "nutritional fact table" can be written in English and French. There are many formats for use on a food package. The standard vertical format is considered for use before horizontal and linear formats. The NFT should not occupy more than 15% of the physical package available display area.

11.2.3.6. China

The Chinese Ministry of Health released the National Food Safety Standard for Nutrition Labeling of Prepackaged Foods (GB 28050-2011) in 2011. The core nutrients that must be on a label are protein, fat, carbohydrate and sodium in 100g/100 ml and energy in kJ.

11.3. Food Meant for Specific Age Group & Convalescing People

The term "convalescent" refers to a person who is recovering from illness or treatment. After recovering from an illness, the body needs to rest and repair. Cells need to regenerate and toxins need to be flushed out. Many illnesses are combatted with strong antibiotics. Sometimes these medications cause severe damage especially to liver and kidney cells.

In medical terms, the nature of illness determines the convalescent from a specific illness, the duration of convalescence, the treatment and the diagnosis. Based on the nature of illness, the convalescent people are categorized into three types.

1. Convalescence from accidents and general febrile illnesses
2. Convalescence from illness and disease affecting the alimentary tract
3. Convalescence from diseases affecting other systems.

11.3.1. Convalescence from Accidents and General Febrile Illnesses

The natural resistance of people collapses before a virulent infection. It includes conditions like exanthemata (skin rash accompanied with fever), common colds, influenza, tonsillitis, etc. Once the infection is over, exhausted cells of the body begin to resume their normal functions. We have to nourish the returning strength without overloading or poisoning the weakened system.

In the earliest phase of convalescence, "milk" forms the basic diet, in the form of coffee and milk, cereals and milk, and milk puddings with cream. Eggs, fish and meat are introduced gradually in small quantities, as protein source. Simple proteins of white fish, chicken, mutton and beef can be given before providing the less digestible "fatty" proteins like pork, duck, goose and salmon. Simple "cereals" form the principal carbohydrate source. Flour, oatmeal, and rice are given as toast, bread, milk puddings and plain cakes, but not mixed with large quantities of fat, fruit and sugar as rich puddings, cakes and pastry. Well cooked vegetables can be given. Fruits is the last option of the foods. A little orange juice may be given occasionally. Sugars should not be encouraged. The amount of fluid intake is left to persons own discretion. Water, milk, tea, coffee, meat extracts and barley water are fluids for the convalescent.

11.3.2. Convalescence from Diseases of the Alimentary Tract

These diseases are divided into two groups – non-specific diseases and specific infections. The non-specific diseases are gastric and duodenal ulcer, cholecystitis and jaundice, appendicitis, diverticulitis. The specific infections are gastro-enteritis, typhoid fever, ulcerative colitis and dysentery.

11.3.2.1. Diet for Group of Non-specific Diseases

"Milk" is the basic diet given in the form of coffee and milk, cereals and milk, and milk puddings. Eggs and fish are the protein in the early stage; poultry and meat towards the end of convalescence. Proteins of ox, sheep, fowl and white fish are better than pork and salmon. Meat should be boiled, grilled or roasted, but should not be served with additional fat. Twice cooked meat or tinner or potted foods, spiced meats, sausages, etc should not be given. Shell fish should be avoided. Butter, cream, cheese and olive oil are given. Large quantities of animal fat or meats fried in fat should not be given. All plain cereals can be given as well-cooked starch of wheat, rice and oatmeal. Cereals with milk or small quantities of butter and sugar can be given. Bread, biscuit to simple cakes and puddings are included, but pastries, fruit cakes and puddings rich in fat, sugar and fruit should be avoided. Well cooked vegetables are given in very small quantities. No raw vegetables or salads and no vegetables containing large quantities of cellulose until the end of convalescence. Fruits should be used sparingly. Sugar may be used as flavourings but no concentrated sugar as sweet, ice-cream or heavily sweetened puddings or drinks allowed. Gifts of chocolates and sweets are forbidden. Water is the best drink. Tea, coffee, meat extracts, milk, soda water and barley water are followed next.

11.3.2.1. Diet for Group of Specific Diseases

These are diseases due to bacterial poison introduced with the food, except ulcerative colitis. The diet of this convalescence group is mainly carbohydrate – cereals, vegetables and salads, fruits and fruit-juices. Sugar is good. Protein should be avoided and introduced on a trial and error basis. Begin with eggs and then, fish and gradually to meat. In ulcerative colitis, permanent cure comes after a long convalescence.

11.3.3. Convalescence from Diseases Affecting other Systems

The diet of this group is mostly a continuance of that prescribed during actual stages of the illness e.g diabetes mellitus and diseases of the blood. Others include renal and cardiovascular diseases. In the treatment of acute nephritis, protein is withdrawn and animal protein is slowly re-introduced into the diet with careful watch on urine. Certain cases pass from an acute to a chronic phase, and albumin in the urine never completely clears up. In such patients, proteins should be withdrawn. In cardiovascular disease, protein must be reduced in the diet during acute phase, and the same restriction be maintained during convalescence. Alcohol should be strictly withdrawn.

11.3.4. Diet for the Aged People

Aged people are vulnerable to nutritional problems due to age related changes in their body, as they have impaired physiological and anatomical functions. The possible nutritional issues in old age are procurement and prepared of foods, psycho-social, digestion, nutrient absorption, renal problems, memory loss, sensory changes and physical problems like weakness, gouty arthritis and painful joints.

An aged person requires less energy than a younger individual due to reductions in muscle mass and physical activity. Some daily requirements for aged people differ from those of normal adults. For example, to reduce the risk for age related bone loss and fracture, the requirement for vitamin D is increased from 200 IU/day to 400 in individuals of 51–70 years of age and to 600 IU/day for those over 70 years of age. The iron intakes reduce from 18 mg per day in women aged 19–50 to 8 mg/day after age 50, due to better iron conservation and decreased losses in postmenopausal women.

Some aged people find difficulty in getting adequate nutrition because of age or disease related impairments in chewing, swallowing, digesting and absorbing nutrients. Their nutrient status is affected by decreased production of chemicals to digest food (digestive enzymes), changes in the cells of the bowel surface and drug–nutrient interactions. Some aged people demonstrate selenium deficiency, a mineral important for immune function. Impaired immune function affects susceptibility to infections and tumours (malignancies). Vitamin B6 helps to boost selenium levels, so a higher intake for people aged 51–70 is recommended. A very high dose supplementation (eg. zinc) gives opposite effect and result in immune-suppression.

There is a reversion to simplicity in diet with advancement of age. Old people cannot stand starvation. Breakfast may include cereals and cream, roll and butter with marmalade, idli with sambar/chutney and occasionally an egg. Tea or coffee to drink. Lunch is the principal meal with a little fish or meat with well-cooked vegetables, milk pudding or boiled pudding, cheese and butter. Water, beer or wine to drink. Tea time with bread, butter, plain cakes and tea. Supper with a glass of hot milk with biscuits. The diet with an adequate calorie value as simple foods is suitable for aged people.

References

Codex Alimentarius Commission 2018. General Standard for the Labelling of Prepackaged Foods CXS 1-1985 Revised in 2018

Codex Alimentarius Commission 2018. Guidelines on Nutrition Labelling. CXG-2-1985 Revised in 2021

Commission Regulation (EC) 1924/2006 of 20 December 2006 on Nutrition and Health Claims Made on Foods. Off. J. European Union. 2006;404:9–25.

General Rules for Nutrition Labeling of pre-packaged foods. USDA Foreign Agriculture Service https://www.fda.gov/food/new-nutrition-facts-label/how-understand-and- use-nutrition-facts-label

Grunert G. 2013, Nutrition Labeling. Encyclopeda of Human Nutrition (3rd Edition). Aarthus University, Aarthus, Denmark.pp 315-319.

Hoadley J.E. 2017. Chapeter 16 – US regulation of food label. In Developing New Functional Food and Nutraceutical Products. EAS Consulting Group, Alexandria, VA, USA. pp 291-308.

Porter D.V. and Earl R.O. 1990. Nutrition Labeling - Issues and Directions for the 1990s, National

12

Seafood Safety Authenticity and Traceability

Introduction

Seafood provides about 25% of the global intake of animal protein and fetches a huge economy for developing nations. Many challenges concerning food safety and quality exist due to globalization and liberalization of the global fish trade. International food safety regulatory bodies such as WTO and CAC through SPS and TBT Agreements work on ensuring food safety and quality. Adoption of risk-based analysis and implementation of the HACCP system are encouraged to maintain high seafood safety standards. International regulatory bodies set maximum residual limits for various environmental pollutants and contaminants to ensure consumer safety. Consumer education programs play important role to enhance awareness and improve transparency in the seafood supply chain. Internationally traded seafood products therefore need to meet the standards prescribed for food safety and quality requirements. In case of non-conformance, they could be detained, rejected and destroyed.

The global demand on seafood quality and safety requirement and the complexities of seafood supply chain make difficulty to ensure authenticity and traceability of seaood products in the market. Most seafood is exported from the country where it is caught while part of it via. a third country. Seafood pass-through complex supply chains involving, capture/culture, transshipment, processing, packaging, wholesale, retailing, and consumer. The global trade of seafood has led to a significant concern associated with seafood fraud.

12.1. Seafood Fraud

Seafood fraud is the practice of providing misleading information to the consumers about their product for financial gain. There are nine types of seafood fraud, which include species substitution, fishery substitution, IUU substitution, species adulteration, chain of custody abuse, catch method fraud, undeclared product extension, modern day slavery, and animal welfare

12.1.1. Species Substitution

Species substitution means substituting an inferior low value species for a superior high value fish species for financial gain, to meet market demand and to evade border controls. For instance, mislabeling of Asian catfish as Grouper to evade high tariff and selling at exorbitant price. Species substitution occurs at various stages in the seafood supply chain. In the global market, it ranges from 25-50% based on the retail market surveys. Seafood mislabeling was first evidenced in 1915 when the excess catches of shark were sold as swordfish.

A variety of market names exist for a given fish species within a single jurisdiction. There is a lack of harmonization in the market names. The USFDA lists over 60 species from seven different genera as Groupers. For instance, the addition of "Chilean seabass" is an acceptable market name for toothfish (*Dissostichus*). Various *Sebastes* spp, are marketed as "rockfish" while "red snapper" is reserved for *Lutjanus campechanus*. Certain rockfish are permitted to be marketed as "Pacific red snapper" or as various snapper names.

Currently, the most commonly substituted seafood products are Asian catfish (*Pangasius*), hake, farmed salmon, shark/rays, tilapia, cod, snapper, and escolar/oilfish. Red snapper, wild salmon and Atlantic cod are implicated in greater number of mislabeling. Tuna (84%) are substituted with escolar, a species that cause serious digestive issues. Table 12.1 gives some of the commonly substituted fish in retail outlets.

Table 12.1. Most commonly substituted fish species in retail outlets

S.No.	Species in label	Commonly Substituted Species
1.	Chilean seabass	Antarctic toothfish
2.	Alaskan/Pacific cod	Pangasius (Asian "catfish"), Atlantic cod, threadfin slickhead, tilapia
3.	Atlantic cod	Pacific cod, white hake
4.	Grouper	Pangasius (Asian "catfish"), king mackerel, whitefin weakfish
5.	Alaskan/Pacific halibut	Atlantic halibut, blueline tilefish
6.	Wild Salmon (king and sockeye)	Farmed Atlantic salmon
7.	Seabass	Antarctic toothfish, Patagonian toothfish
8.	Snapper	Giltheaded seabream, madai, tilapia, Pacific ocean perch, widow rockfish, yellowtail rockfish
9.	Red snapper	Caribbean red snapper, crimson snapper, spotted rose snapper, Pacific ocean perch, yellowtail rockfish, gilt-headed seabream, madai, tilapia, white bass
10.	Lemon sole	Blackback flounder, summer flounder, flathead sole, yellowfin sole
11.	White tuna	Escolar
12.	King scallops	Japanese scallops

Seafood substitution poses a significant public health threat due to allergens, high toxicity or contaminants. For instance, pufferfish (*Lagocephalus scleratus*) contains a neurotoxin (tetradotoxin) is often mislabeled as monkfish to evade import restrictions and for economic gains. A survey reports more than 50% of fish species substituted can pose health risks to consumers. The highest level of substitution occurs at retail outlets including restaurants and grocery stores.

12.1.2. Fishery Substitution

Fishery substitution means the substitution of a product from a superior fishery with an inferior one. For instance, shellfish harvested from a Class C with that of Class A bed. This causes public health risk as the substituted one is harvested from polluted waterbodies with high numbers of *E.coli*. Fishery substitution is prohibited in the EU, as it requires the catch area or country of origin. The consequences of fishery substitution are economic deception, industry issues, consumer rights, and public health risks. Fishery substitution is more commonly done by the distributors, middlemen, and food service companies, to avoid foreign duties.

12.1.3. IUU Substitution

The IUU fishing includes capture of fish in areas closed for harvest, undersized fish harvest, illegal setting of traps, misreporting of catch as low value species (i.e. pink salmon reported as chum salmon) and use of unlicensed vessels. It is reported that 10% of the total fish imported to the EU originate are from IUU fishing. The IUU fishing is facilitated by nations, by providing flags of convenience, and relaxing import and export regulations. Illegal fishing vessels supplies mixed fish catch due to transshipment and sale at sea. The traceability issues are more when companies own fishing vessels in another country or register it under another national company or flag.

The global nature of seafood supply chain is that products are transported via. intermediary countries after post-harvest processing. This provides multiple opportunities for mixing of legal and illegal sourced fish. Illegal fish launders in processing countries and enters international trade as a legal product. For instance, Russian vessels harvest illegal crabs and transship them to flags of convenience (Japan or Korea), which undergo processing with counterfeit Certificate of Origin and Health for shipping to China. Illegal and legal crab get mixed up and exported to developed nations from China. This kind of fraud goes undetected due to false documentation, repacking and obfuscation of traceability. The IUU supplies are more in countries that lack effective traceability systems such as China, Japan and Canada, which are the major seafood exporters and importers.

The EU enhanced the legislation and stipulated catch documentation requirements for all imported and exported seafood. A third country card process imposes import restrictions on countries doing IUU fishing and penalties for those engage or support IUU fishing. The EU has stringent traceability and mandatory labelling requirements from catch or harvest to the retail level to decrease seafood fraud. The limitations in these regulations are the exemption for most processed or prepared food (cooked, steamed, breaded, fried or marinated), several types of invertebrates (jelly fish) and canned seafood; and also the food service sector of the supply chain (restaurants, take-away, school, and catering units).

12.1.4. Species Adulteration

Species adulteration is the addition of undeclared, non-specified species to a primary processed raw material. For instance, addition of a low value fish (Saithe) to a high value frozen block of Atlantic cod. The motivation for this fraud is exploitation of the price difference between different species or introduction of dubious or prohibited origin in the supply chain. The seafood can be adulterated with additives to increase their production or food quality or to hide other types of fraud. The presence of these additives or chemicals may be detrimental to human health or not permitted treatments. Prolonged consumption can cause inflammation, ulceration, and necrosis in the linings of the mouth, throat and gastrointestinal tract leading to cancers. Bleaching of seafood in a mixture of water and hydrogen peroxide to make fish seem whiter, fresher and more attractive is done in Italy. Formalin is used in Southeast Asian countries to preserve the fish fresh with stiff texture, which are all violation of law. Species adulteration is reported throughout the supply chain.

12.1.5. Chain of Custody Abuse

Chain of custody is the chronological documentation containing the seizure, custody, control, transfer, analysis and disposition of evidence, which is a part of traceability and certification processes. The falsification of documentation allows financial gain from premium credence claims (consumer value issues) and provenance claims (packaging). The chain of custody abuse is more due to the global and complex nature of seafood supply chain comprising of many actors and activities, which range from fishermen and aquaculture farmers, transshipments, distributors, warehouse owners, wholesalers, retailers, brokers, large trading companies. They tamper with the traceability documentation through the supply chain. Import health authorities, food safety officers or custom authorities do have full access to the information pertaining to the origin, transshipment points, and other activities associated with the product. Therefore, traceability of seafood throughout the supply chains is inconsistent and challenging. Incidences of chain of custody abuse are reported. For instance, a seafood firm sold several million dollars' worth of expired products in the past three years to restaurants and traditional markets in Taiwan by removing expiry dates from packaging and re-selling old shrimp to mask the freshness.

12.1.6. Catch Method Fraud

Catch method fraud involves the mislabeling of the type of production or harvesting method for financial gain. For instance, line caught fish attract a higher price than trawled fish. Likewise, the customers prefer a higher priced wild species over the farmed species. A UK study identified that 15% of salmon, 11% of seabream and 10% of seabass as farmed, which was claimed as wild. This fraud is performed mainly at the processing and manufacturing supply chain by agents, middlemen or final retain customer.

12.1.7. Undeclared Product Extension

Undeclared product extension means the use of technology by processors to increase the perceived weight of the seafood content for economical gain. The practices include over treatment (i.e. over breading or over-glazing), soaking fish in a brine solution, injection of undeclared chemical additive to increase the water holding capacity of the muscle or injection of fish by-products (minced up and blended) back into the fillet to bring up the weight. For instance, in China, injection of gelatin like chemicals derived from animal skins and bones into prawns and shrimps is reported. The chemical resembles like natural colour and increase the weight by 20-30%, which allow the wholesalers to increase their price. The long-term injection of unknown chemicals and industrial substances can cause public health threat.

12.1.8. Modern Day Slavery

Modern day slavery is a situation when someone is forced to work by ownership or control of an employer through mental or physical threat or abuse, dehumanized or physically constrained of free movement, which is illegal. Foods produced using high standards of human welfare is considered a quality attribute by consumers, which enhances brand reputation, competitive advantage and price premium. Therefore, fraudulent claims and undetected modern day slavery in the food supply chain deceives the consumer and reduces financial gains. Seafood supply chain is accused of exhibiting modern day slavery and expose the employees to a greater risk of injury, death and human right abuses. For instance, between 1948 to 2008, 1039 fatalities have been reported in UK fishing vessels that were unstable, overloaded and unseaworthy. Likewise, in China 23 out of 28 illegal immigrants died in a Bay in 2004 by a rising tide while collecting cockles due to minimal employee safety. The UK based Serious Organized Crime Agency (SOCA) reported the victims of illegal, bonded and forced labour in fisheries, who are not paid their wages and experiencing poor working conditions. Aquaculture sector offers a safer occupation than working in wild capture. However, fatality rates are also reported in Scottish aquaculture industry.

It is very difficult to track a product of a particular producer as the seafood supply chain is more complex and internationalized and hence, it is a challenge to certify social welfare and ethics. Abuse of human welfare occurs when customers give their supplier a large order beyond their capacity, forcing the supplier to sub-contract work to workers, who are not regulated by the same standards. When buyers negotiate for low prices, there is an increased likelihood for the use of forced labour to reduce the price. A number of Government and independent initiatives have developed standards to promote responsible supply chain activities, particularly labour rights, health and safety, environment and business ethics. They are Modern Slavery Act 2015, Ethical Trading Initiative 2016, and Marine Stewardship Council 2016. The companies should take responsibility to ensure that no forced labour is used throughout the seafood supply chain.

12.1.9. Animal Welfare

A new unconventional fraud identified relates to the sale of food with valuable animal welfare marketing claims such as Animal Friendly. Animal welfare is an important aspect for many consumers. It is quiet challenging to prove that a seafood product meets all the requirements of welfare, including a suitable environment and diets, to exhibit normal behaviour patterns and to housed apart from other animals and to be protected from unnecessary pain, injury, suffering and disease. Animal welfare of seafood is undermined in the supply chain from primary production to its slaughter.

In the wild capture fisheries, welfare becomes an issue from the moment the fish encounter the fishing gear. Animals either die as a consequence of the harvesting process or enter the fishing vessel alive, especially in the case of high value fish. The time of death can vary from minutes to hours to days and each animal experience different gear specific trauma. In aquaculture, welfare is an issue through the supply chain to the point of slaughter. Stocking density, water quality, diet, feeding techniques, husbandry practices and management procedures also affect welfare prior to death. More humane slaughter methods are utilized for fish. For instance, automated percussive systems or electric stunning systems are developed to stun fish immediately allowing them to die without regaining consciousness. Crustaceans encounter uncomfortable experiences during their capture and death, particularly during their live storage and transport. Electric stunning prior to processing reduces unnecessary abuse to their welfare.

Seafood species from aquaculture or wild fisheries may be exposed to unnecessary pain through their environment, capture, harvest, transport and slaughter. Welfare claims are used as a tactical marketing ploy to gain from the good publicity of the labelling. For instance, one case of canned tuna was labelled as "dolphin safe" despite the fact that skipjack tuna is not implicated in the dolphin bycatch problem. However, the current regulation is limiting the way fish are handled or killed. The active certification schemes are taking care of these ethical issues.

The key part of any regulation is the accurate identification of fish being caught, imported and sold to consumers at retail outlets or restaurants. The identification of seafood products when processed and packaged is difficult because of the lack of morphological characteristics. To overcome these, new DNA based tools are used for checking the accuracy of seafood labeling by validating the identity of fish species in the package label with market name in which it is sold. DNA barcoding is the most common method used for species identification in market surveys. Visual identification by simply comparing package label with menu is also used to identify fraud in restaurants. Market surveys indicate higher levels of mislabeling in processed products when compared to whole fish.

12.2. Traceability

Traceability is defined by the Codex Alimentarius Commission as "the ability to follow the movement of a food through specified stage(s) of production, processing and distribution". Traceability provides knowledge regarding the identity, history and source of a product, or of ingredients contained within a product. It also provides knowledge regarding the destination of a product. It is therefore an information management tool.

In the fishery sector, traceability is used in relation to three aspects:

1. **Food safety** i.e. to ensure that products and ingredients come from origins that meet food safety conditions
2. **Application of tariffs and quota tariffs** i.e to ensure that appropriate rates of duty are applied
3. **Sustainability** i.e. to ensure that the fish is derived from sustainable sources, such as from vessels that follow conservation rules (e.g. for catch certification schemes)

Traceability is important as it allows the operators to guarantee the safe origin of their product, and to take appropriate actions (such as withdrawal or recall), if it is note safe.

12.2.1. Need for Traceability

12.2.1.1. Meeting Specific Regulatory Food Safety Conditions

Traceability is a national or export requirement set out in regulations for supply of seafood by nations. In the EU, there is a specific requirement expressed under Article 18 of Regulation (EC) No 178/2002 of the European Parliament and of the Council on 28 January2002. It requires that all food business operators, feed producers and primary producers of animals to have in place a "one-up and one-down" traceability system, which is also applicable for operators in third countries, which supply the EU with products.

The Code of Federal Regulations requires importers to the U.S. to maintain records to identify the immediate sources of their foods, which has to be maintained for at least two years. The Bioterrorism Act of 2002 also requires domestic and foreign facilities that manufacture, process, pack or import food in the United States to register with the USFDA. The Section 204 of the US Food Safety Modernization Act 2011 requires the USFDA to establish a system that will enhance its ability to track and trace both domestic and imported foods. The FDA is recommended to establish a uniform set of record-keeping requirements for all foods. It is also recommended to require food business operators at all levels of the food chain to develop, document, and implement a product tracing plan.

12.2.1.2 Voluntary Measures

Apart from regulatory requirements, several voluntary certification schemes related to production conditions for fishery products also require traceability systems.

- Marine Stewardship Council (MSC) certification for sustainable fisheries,
- Global GAP (Good Agricultural Practices) Aquaculture Standard Version 5.
- ISO 22000:2005 Food safety management systems

Traceability is also referred to as **chain of custody**. The product is to be displayed with the certification logo stating that it is derived from the fishery or aquaculture operation meeting the standards.

Traceability systems also make good business sense for responsible business operators, as it makes management of food safety incidents easier. Systems such as trace-back, trace-forward, product withdrawal and product recall, cannot function without provisions for traceability. So, traceability must be operated in conjunction with GMP and HAACP to deliver safe food.

12.2.1.3 Complexity of Supply Chains

The complex seafood supply chain makes traceability a considerable challenge to meet all requirements, as it consists of numerous separate business operators. An example of a supply chain for aquaculture products is shown in Fig. 12.1.

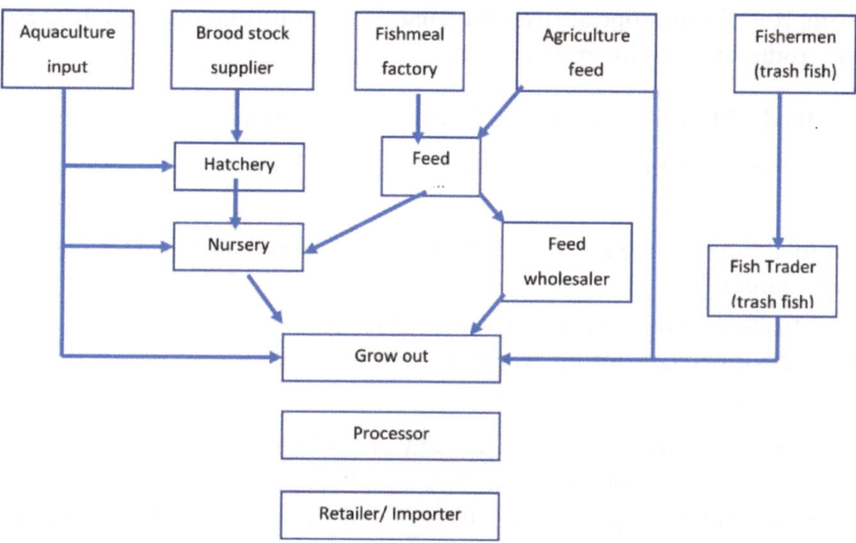

Fig. 12.1. A typical aquaculture food supply chain

In the modern fisheries and aquaculture business, supply chain is complicated by the spatial dimensions, where inputs (feed, fish etc.) and outputs (final products) are traded internationally. Such complexity occurs even within vertical integrated businesses i.e. in multi-site operations. So, to ensure a sustainable business, compliance with SPS measures and traceability systems are required to trade across international borders.

12.2.2. Key Elements of a Traceability System

12.2.2.1. Objectives

The ISO 22005:2007 standard "Traceability in the feed and food chain – General principles and basic requirements for system design and implementation" explains the principles and requirements for the design and implementation of a feed and food traceability system. This standard is one of the ISO 22000 series of food safety standards that sets out the conditions for certification of a traceability system.

The objectives of the traceability system are

1. To trace the flow of materials (feed, food, their ingredients and packaging)
2. To identify necessary documentation and tracking for each stage of production
3. To ensure adequate coordination between the different parties involved
4. To improve communication among the involved parties
5. To improve the appropriate use and reliability of information, effectiveness, and productivity of the organization.

12.2.2.2 Basic Characteristics of a Traceability System

The basic characteristics of a traceability system within a business operation are:

1. Identification of incoming products (or raw material and ingredients) and their sources
2. Identification and recording of information on activities linked to these products or batches during processing and storage
3. Identification of outgoing products, and their destinations

The traceability system comprises a **data recording** and **retrieval system** that links these steps. The first and last activities require gathering data at the interface with suppliers and customers and the middle activity requires tracking of the operations within the business operation, by means of labelling, separation of different batches, and recording when batches are split or mixed.

12.2.2.3 Data on Inputs

The data on inputs identifies an exclusive list of suppliers of materials and ingredients, which could form part of the final product. A batch record is prepared on receipt of the physical inputs, which records the supplier, the date, the description of the product and any batch codes contained within the received consignments.

The operator then applies his own codes or identifiers for **internal purposes**. An identifying code is applied to each batch of raw material on reception. Supply information such as name of supplier, date, quantity, species and other characteristics, and any supplier batch codes, is recorded against this code.

In capture fisheries, the material inputs are water and ice. The vessel operator should record information relating to the supply of these substances. In aquaculture, there are several inputs, notably eggs or juveniles, feed materials (including additives and supplements), and veterinary medicines. In processing operations, inputs include fish, along with other ingredients and additives. Packaging materials are also included as an input in food traceability systems due to the potential contamination of the product.

12.2.2.4. Data on Production

In capture fisheries, a record of the vessel (i.e. registration number), date, fishing location, gear used, and time of capture is made. Other useful information such as ambient temperature and seawater temperature that can impact on food safety conditions can also be recorded. **Batch separation** is practiced to avoiding mixing of old and new catches, which is maintained until discharge.

Aquaculture operators record all activities involved in the production of the fish, which include pond or cage location/number, dates, and quantities of application of feed (indicating the batch numbers of the feed), along with other treatments (supplements, veterinary medicines).

In fish processing steps, all treatments and associated data (e.g. HACCP records) applied to that batch are registered to that code. Associated data include storage location and conditions, date and shift of work, along with production yields.

12.2.2.5 Batch Separation, Mixing and Combination

It is essential to ensure that there is effective batch separation of batches during the production process to ensure traceability. In aquaculture operations, fish are grown in separate ponds, cages, or tanks, which define effective separation between batches. In processing lines, batch separation requires processing different batches on different lines, or at different times. Fish processors use labelled or colour-coded fish boxes and tags to identify different batches during the process.

However, in fish processing operation, batches of fish are often mixed (i.e. after grading in aquaculture), or made as a composite product (i.e. in processing). These actions should be recorded, because there is a loss of integrity in the traceability data. By keeping a record of batch mixing, there are possibilities to identify the precise origin of a final product. It is necessary to split a batch and proceed with different processes for separate parts. The two new batches generated are allocated with new codes and should be related to the original code.

In the fishery sector, where supplies are often derived from large numbers of small-scale fishing vessels, it is an essential feature of the business that batches are mixed, and it is not always be possible to maintain traceability to a single fishing vessel. However, efforts can be made to preserve the integrity of traceability data, at least to the level of the landing site.

12.2.2.6 Data on Outputs

The operator must maintain a record of outputs with sufficient information to link the final product to all of the data collected on inputs and processing.

In simple operations with processor and one supplier, supplying one batch of a single species per day, the date of production is sufficient. In more complex operation, a batch coding system is devised and applied to allow the process batch to be identified. The code is affixed to the product, or to the packing, or contained in the associated documentation to be a reference for the receiver. This provides the essential data link to the next operator in the supply chain. The receiver is identified as part of the data collected on the outputs. There is no need that batch codes are understandable by the receiver, as it can be an internal code unique to the business operator. This is referred to as an internal traceability system.

12.2.2.7 Recording, Storage and Retrieval of Data

All the data collected on inputs, process, and outputs are collated by the business operator and stored in a **data retrieval system**. This system provides a full history of any batch of product. Two core tasks are performed by the data system:

The system is able to identify the business operators who received any product containing that input, with a reference to a specific input

The system is able to identify the business operators who supplied, and the inputs contained in that output, with a reference to a specific output

These tasks are the core of all traceability systems. They represent the basis of the "one-up, one-down" concept of minimum traceability, in which an operator is able to identify the supplier and the receiver of any given product, which has been in his/her possession. If the business operators provide written invoices, the information is recorded on the invoice. If not, the form in the box provides a means of capturing transaction data in a small-scale fishery or aquaculture business. Such a form can be used by fishery business operators to record their transactions. (Table 12.2). Two copies are made of the form, and supplier and receiver each keep one for their records.

For storage of data, small operators, or operators with limited supply and marketing channels use paper-based systems. Traceability systems are amenable to computerization, and many software packages are also available. Some of them are dedicated to fishery sector. The minimum period specified for keeping the records depends partly on the expected shelf life of the product during its subsequent distribution. The records for long shelf products (e.g. canned fishery products) need to be kept for several years (typically 2 to 5 years), whereas for fresh fishery products, the period is a matter of months.

Table 12.2. Model forms for maintaining traceability by Supplier and Receiver

Supplier Name:	**Supplier Registration No:**
Supplier Contacts:	

Batch No./ Identification	**Description**	**Quantity**

Receiver Name:	Receiver Registration No:
Receiver Contacts:	

Supplier Representative	**Receiver Representative**
Name : Signature : Date:	Name : Signature : Date:

12.2.2.8 Internal and External Traceability

12.2.2.8.1. Internal Traceability

Internal traceability system applies to a single business operator, where the system is internal to that operator (Fig.12.2). Each operator in the chain develops such a system and applies it. The two core tasks are performed and the conditions are placed for a "one-up, one-down" traceability.

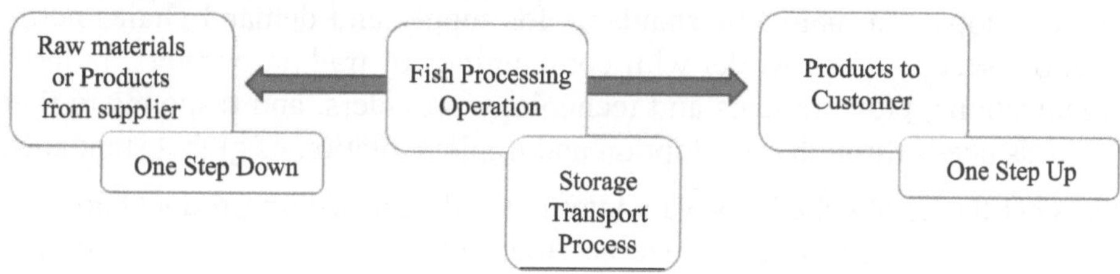

Fig. 12.2. One stage in the Seafood Supply Chain

With internal traceability, each operator within a supply chain maintains their own independent system. They are obliged to exchange information with their suppliers (i.e. batch identifiers). Thus, the operators who receive product will also receive an associated code, which identifies the fishing vessel supplying the product.

Data regarding suppliers and customers is a sensitive information for any business. With internal traceability, this data resides within the business operators' system ie.for tracing the cause and origin of a food safety problem or non-compliance identified later in the supply chain. The operator thus applies the system to make the information available on demand by the competent authority.

With internal traceability, each operator in the food chain is responsible for the operations under their control, and there is no requirement for whole chain traceability.

12.2.2.8.2. External Traceability

An external traceability system requires all traceable items to be uniquely identified, and information on them to be shared between all of the participants within the supply chain. Thus, an operator at the consumer end of the chain is capable to identify, not only his direct supplier, but also their suppliers.

To maintain this system a standardized approach to traceability identifiers is required. Such systems are imposed by major buyers i.e. multiple retail operators, who receive an own-brand product from several suppliers and require them to adopt an in-house traceability system, as part of the conditions of sale.

A range of international standards is developed by ISO, that apply a systematic approach to operation of an external traceability system.

12.2.3.Traceability Tools

12.2.3.1. GS1 Barcode Systems

Global Standard 1 (GS1) is a neutral, not-for-profit, international organization that develops and maintains standards for supply and demand chains across multiple sectors. GS1 works with communities of trading partners, industry organizations, Governments and technology providers, and responds to their business needs through the adoption and implementation of global standards.

The GS1 has established a private international standard for product barcodes, and has developed specific external traceability systems for the different segments of the food industry, including the fishery sector. The GS1 principles of product coding are

1. **Identify:** The globally recognizable identification of trade items, products and locations using GS1 identification standards (i.e. unique numbering standards).

2. **Capture:** Capturing the unique identification of trade items, products and locations using automatic data capturing technologies (i.e. barcode scanning, Radio Frequency Identifying Technologies, RFID).
3. **Share:** Sharing the information about trade items, products and locations internally within the business and with trading partners, in a standardized manner using computer networks and messaging standards.
4. **Use:** Applying the foundational standards to business processes

The system defines GS1 Application Identifiers in the form of database fields, which is expressed in a numerical code. These can be printed as a barcode, or coded into a RFID or other system. A barcode is an optical machine-readable representation of data relating to the object to which it is attached. Barcodes systematically represent data by varying the widths and spacing of parallel lines (1D) or rectangles, dots, hexagons and other geometric patterns in two dimensions (2D). Barcodes are originally scanned by special optical scanners or by using interpretive software available on smartphones.

Product traceability initiatives use a Global Trade Item Number (GTIN) to achieve traceability. A GTIN includes a GS1 company prefix and a unique item reference number compatible with Universal Product Code bar codes, and RFID or human readable codes.

12.2.3.2. Radio-frequency Identification (RFID) System

A number of operators in the fishery sector employ RFID, which uses electromagnetic fields to automatically identify and track tags attached to objects. The tags are written with electronically stored information (i.e. traceability code). **Passive tags** collect energy from a nearby RFID reader's interrogating radio waves. **Active tags** require a local power source such as a battery and operate at hundreds of meters from the RFID reader.

RFID offers advantages over manual systems or bar codes. Unlike a barcode, the tag need not be within the line of sight of the reader, so it can be embedded in the tracked object such as fish box or pallet. RFID tags can be read hundreds at a time.

12.2.3.3. TraceFish Standards

TraceFish is the short title for the "Traceability of Fish Products", an EU funded project (2000) coordinated by the Norwegian Institute of Fisheries and Aquaculture. The project developed three consensuses- based standards for recording and exchange of traceability information in the fishery supply chain:

1. Farmed Fish Traceability Standard
2. Captured Fish Traceability Standard
3. Technical Standard

The standards lay down where, what and how data should be recorded in the farmed and wild caught fish chain for full chain traceability. They identify how modern electronics and software can be used to transmit external traceability data through the chain, and the standards to be used to successfully obtain the data. All three TraceFish standards are publicly available. The two fish industry standards are sold and distributed through CEN, the European Committee for Standardization. The third data standard is TraceCore, an open-source XML software package. It provides a standardized way to transfer data on traceability, with standard names, reference values, terms, measurements and values. It allows suppliers to deliver information in one known format, and for receivers to be able to read it in the same format.

12.2.3.4. Proprietary Software Applications

A number of software applications have been developed specifically to meet the needs of fishery business operators for traceability. They incorporate a different data logging, such as HACCP variables, cold chain integrity, etc., which are largely automated to reduce the cost of data collection and to remove the scope for human error. Some of the most used packages are Trace Register, Shellcatch, Scoring Ag and Trace Tracker.

12.2.3.5. Recall and Withdrawal Procedures

The value of traceability system comes to force during incident management, i.e. when a food safety problem is detected. The problem could be due to a non-compliance (i.e. HACCP monitoring, or a regulatory testing), or a food poisoning outbreak.

It is the responsibility of the food business operator to follow up the problem, and to identify its cause and origin, which is undertaken voluntarily, or it is forced on the business operator under legal powers of the Competent Authority.

Many food safety problems originate elsewhere in the supply chain other than the point where they are detected, and there is a need to undertake a trace-back operation, to allow checks, such as inspections, checks on HACCP records, sampling and testing, to be made on the food safety conditions at each stage, until the source of the problem is identified.

Once the source and nature of the problem is understood, there is a need to consider that the deficiency has resulted in unsafe products distributed via other supply chains. There is also a need to trace forward, to identify all operators who have received the unsafe food, and inform them to withdraw or recall the product. The approach to trace-back and trace-forward is given in Fig. 12.3.

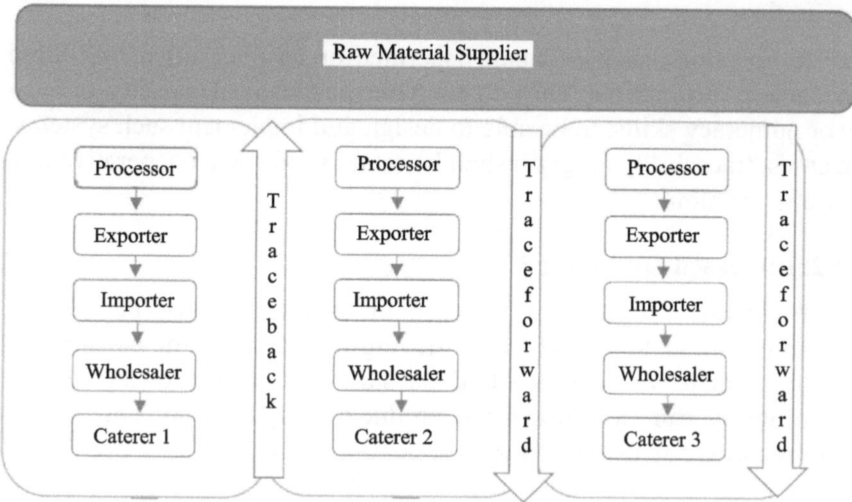

Fig. 12.3. Trace back and trace forward for withdrawal and recall of unsafe food

12.2.3.5.1. Recall Plan

A **recall plan** is implemented when the food has reached the consumer, but may not have been consumed. The aim of the recall plan is to inform the consumer that the product should not be consumed and that they should return it to where they bought it, and seek a replacement or refund. It is a major work of a food safety competent authority to ensure that food recalls are properly designed and implemented

The plans should be written. Most business operators cooperate and implement the plan on instruction. Competent Authorities have the legal power to require product withdrawal or recall. Product recalls should be publicized by the Competent Authority and widely circulated in the press to make sure that consumers are aware of it. Normally such plans are only required of processors and distributors. Primary producers (fishers and aquaculture operators) do not have the capacity, but should provide traceability information, when requested.

12.2.4. Costs and Benefits of Traceability

12.2.4.1. Costs

12.2.4.1.1. Additional Investment

The traceability system should be considered as a capital investment, in terms of system design, preparation of forms, and recruitment and training. For that, additional staff and required and purchase of any special equipment or software are required. The implementation of a traceability system requires more space for storage of raw material and final products, to ensure effective batch separation.

Small scale operators, such as small fishermen or farmers, often find these investments to be prohibitive. They do not have the technical capacity, or even literacy or numeracy skills, to be able to design and implement such systems. In these cases, traceability program shall be established via a cooperative or a producer organization.

12.2.4.1.2. Increased Operating Costs

Traceability incurs a significant increase in operating costs, because of recording information, storing it and retrieving it, which is time consuming. The operators need to consider the additional staff time required on an ongoing basis. Traceability can slow down the production process, not only in the keeping of records, but also in terms of separation of batches.

12.2.4.2. Benefits

12.2.4.2.1. Damage limitation in Food Safety Failure

Traceability is a tool to limit the impact of a food safety failure in terms of illness or death. It benefits the food business operator who will be obliged to withdraw and destroy all batches, if problem is identified. Testing every unit in a batch is often not feasible. If the affected products cannot be identified and separated, an inspector is obliged to consider any detectable food safety non-compliance as grounds for the batch to be condemned.

12.2.4.2.2. Better Process Controls

Traceability improves stock control and reduces out-of-date product losses, lowers inventory levels, quickens the identification of process and supplier difficulties, and raises the effectiveness of logistics and distribution operations.

Additionally, operators can collect quantitative data on yields associated with specific batch codes. Operators can gain a better understanding of the critical process variables by relating yields to independent variables concerning the process conditions and improve the efficiency of their production processes to get financial benefit.

12.2.4.2.3. More Secure Markets

An improved traceability provides better food safety management and greater guarantees in terms of sustained market access and buyer confidence in the longer run. Improved customer confidence helps with branding and improved brand equity. Traceability is a marketing tool providing the customers with unique information about the product and its origins.

12.2.5. Inspection and Official Control

Traceability is well established as an important food safety requirement encapsulated in legal systems. Inspectors have a responsibility in their official controls of fishery and aquaculture business operators, to check that the traceability obligations exist in the system. Checks on the existence of a traceability system and its implementation, the presence of withdrawal and recall plans are required as part of the routine inspections of the fishery establishment.

The actionable offences leading to launch of a non-compliance procedure are:

- failure to keep traceability records; make false records
- failure to disclose information when lawfully requested
- failure to keep a recall or withdrawal plan
- failure to implement a notice to implement a recall or withdrawal

12.2.6. Traceability Standards and Regulations

There are three main categories of traceability standards and regulations

1. International standards and guidelines
2. Regulatory standards
3. Industry and NGO non-regulatory standards

12.2.6.1. International Standards and Guidelines

International standards and guidelines are developed to provide bestpractices in tracing food products through supply chains. They are standards and guidelines developed by the Regional Fisheries Management Organizations (RFMOs) and Natural Resources Management Inter-Governmental Organizations with the aim to provide guidance to their Member States in dealing with illegal, unreported and unregulated (IUU) fishing.

12.2.6.1.1. The Codex Alimentarius Commission (CAC)

The Codex Alimentarius or "Food Code" was established by FAO and the World Health Organization in 1963 to develop harmonized international food standards, which protect consumer health and promote fair practices in food trade. The Code defines traceability as "the ability to follow the movement of a food through specified stage(s) of production, processing, and distribution" (CAC, 2006). The CAC is recognized by the World Trade Organization as an international reference point for the resolution of disputes concerning food safety and consumer protection. The CAC recognizes that at the international level, methods are not harmonized and are often complicated, thus leading to barriers to trade.

12.2.6.1.2. The International Office of Epizootics (OIE)

The Aquatic Animal Health Code ("Aquatic Code") of World Organization for Animal Health (OIE) or World Organization for Animal Health (WOAH) sets standards for the improvement of aquatic animal health and welfare of farmed fish worldwide and for safe international trade in aquatic animals and their products. The health measure in the aquatic code should be used by the competent authorities of importing and exporting countries for early detection, reporting and control of agents that are pathogenic to aquatic animals and for the prevention of their transfer via international trade, while avoiding unjustified sanitary barriers to trade (OIE, 2005). OIE supports its Member Countries and Territories in implementing animal identification and traceability systems in order to improve the effectiveness of their policies and activities relating to disease prevention and control, animal production, food safety, and the certification of exports. The Aquatic Code emphasizes that traceability should be a demonstration of Government Veterinary Services' capacity to exercise control over all animal health matters, and not a description of the responsibility of private stakeholders in the chain.

12.2.6.1.3. FAO Guidelines

12.2.6.1.3.1. Marine Capture Fisheries – Ecolabelling

The FAO Guidelines (2009) for the ecolabelling of fish and fishery products from marine capture fisheries (2009) summarize several principles that should be observed. by ecolabelling schemes. These guidelines cover the fisheries management system, the status of the target stock, and ecosystem considerations, with the purpose of identifying sustainable fisheries. Paragraph 16 of the guidelines state that the CoC measures designed by the ecolabel "should cover both the tracking/traceability of the product all along the processing, distribution, and marketing chain, as well as the proper tracking of the documentation". In the industry, there is a difference between traceability and ecolabel type CoC, even though the two terms have same meaning.

12.2.6.1.3.2. Aquaculture - Certification

The FAO Technical Guidelines on aquaculture certification provides guidance for the development, organization, and implementation of credible aquaculture certification schemes (FAO, 2011). The guidelines address a range of issues that should be considered relevant for certification in aquaculture such as (i) animal health and welfare (ii) food safety (iii) environmental integrity and (iv) socio-economic aspects associated with aquaculture. These guidelines state the one of the principles of aquaculture certification schemes is that they should include adequate procedures for maintaining CoC and traceability of certified aquaculture practices and processes.

12.2.6.1.4. RFMO catch/ Trade Documentation Schemes

RFMOs are international organizations formed by countries with fishing interests in area beyond national jurisdictions (ABNJs). Some of them manage all the fish stocks found in a specific area, while other focus on particularly highly-migratory species, notably tuna, throughout vast geographical areas. Some RFMOs have a purely advisory role. Many RFMOs have management powers to set catch and fishing effort limits, technical measures, and control obligations (European Commission, 2015). Almost all the global high seas are now covered by at least one RFMO out of the 18 current Fisheries Management Bodies. Of these, five are the tuna RFMOs, which manage fisheries for tuna and other large species such as swordfish and marlin. RFMOs are using catch and trade documentation schemes as part of their fight against IUU fishing. These schemes are important fisheries management tools, but are not designed as traceability system for markets/ consumers.

12.2.6.2. Regulatory Standards

Regulations are set by some countries that are broadly applicable to food products and more specifically to fish products. Regulations are mandatory for export to the European Union (EU), the United States of America (USA) and Japan. There are laws, regulations, and associated enforcement programmes for traceability of fish products. These norms set the minimum traceability requirements for all trading of food products, as well as fish specific requirements for preventing trade of illegally caught fish.

A ranking of countries based on assessment of traceability program was done – Progressive, Moderate and Regressive countries. Progressive country has specific traceability regulations for all commodities e.g. Austria, Belgium, Denmark, Finland, France, Norway, Switzerland, Germany, Ireland, Italy, The Netherlands, Sweden and UK. Moderate country has some or less broad or stringent regulations e.g. Australia, Brazil, Canada, Japan, New Zealand, USA. Regressive country is in development stage with traceability regulations e.g China, India. In many countries, seafood labeling regulations are not very strict in processed products.

12.2.6.2.1. European Union (EU)

EU regulations address seafood fraud and fisheries management. The Common Fisheries Policy (CFP) sets the regulatory framework for the exploitation of the living marine resources. A number of rules and laws are framed and implemented to fight against illegal fishing, and to protect food safety, hygiene, and consumers. Traceability and authentication are pillars of control and enforcement of laws and rules.

The EU legislature works with two different models with respect to food traceability. The model implemented through the application of the General Food Law (EU, 2002) leads to a generic (non-specific) low warranty traceability of the EU food supply chain. A second, more complex model is followed in norms regulating products such as those derived from genetically modified organism (GMOs). Traceability requirements for GMOs in the EU made DNA testing to be integrated into the traceability scheme.

Article 18 of the European Commission Regulation 178/2002 laying down the general principles and requirements of food law, establishing the European Food Safety Authority (EFSA) and laying down procedures in matters of food safety, the legal requirements rely on the "one step back – one step forward" approach to traceability. This approach implies that food business operators (FBOs) have to establish two types of links: a link "supplier-product" (which products supplied from which suppliers) and a link "customer-product" (which products supplied to which customer). Nevertheless, FBOs do not have to identify the immediate customers when they are final customers. The regulations do not establish internal traceability i.e. incoming and outgoing products. There is no requirement for records to identify split or combination of product batches within a business (EC, 2010). Companies know where the ingredients came from and where the products went, but not necessarily which ingredients went into which product.

The current EU regulations that ensure seafood authenticity and traceability are:

1. EC No. 1224/2009 ensuring compliance with the rules of Common Fisheries Policy
2. EC No. 1005/2008 IUU Regulation concerning Illegal, Unreported and Unregulated (IUU) fishing
3. EC No. 1006/2008 fishing authorization regulation for fishing activities of community fishing vessels outside Germany waters and the access of third country vessels to community waters.

The EC No. 1005/2008 establishes a community system to prevent, deter and eliminate IUU fishing. It is the legal base to identify IUU fishing, applied to all fishing vessels. It seeks to ensure full traceability of all marine fishery products traded with the EU. It excludes freshwater fishery products, aquaculture products, and ornamental fish. Chapter III gives the Catch Certification Scheme for importation and exportation of fishery products. The Catch Certificate may be established, validated or submitted by electronic means or be replaced by electronic traceability systems. Regulation (EC) No 1010/2009 provides the regulation for establishing the Catch Certificate.

Common Fisheries Policy Control Regulation specifies control requirements based on established and new technologies, such as vessel detection systems and automatic identification systems. Important elements are:

1. Log book for vessels of 10 m length or more, submit landing and transshipment declaration (electronically for vessels of 12 m or more), prenotification of landings
2. Transshipments only in designated ports and monitoring transshipments by competent authorities
3. Whole chain of production and marketing covered by coherent traceability law (complementary to General Food Law)

EC Regulation No. 1224/2009 – general food law requirements for fishery and aquaculture products state that all lots are traceable at all stages of production, processing and distribution, from catching or harvesting to the retail stage. The fishery and aquaculture products that are marketed within the EU shall contain information concerning the commercial designation, production methods, and the catch area at each stage of the marketing as stipulated in EC No. 1379/2013.

12.2.6.2.2. United States Food and Drug Administration (USFDA)

Species substitution is an economic fraud and misbranding violation under the Federal Food, Drug and Cosmetic Act in USA. There are potential species-related and process-related hazards. The effect of mislabeling through species substitution on the identification of potential species is a species-related hazard. Substituted fish pose health hazards. One frequent substitute for tuna is escolar, which is hard-to-digest oil fish. Some other fish contain unique parasites that can cause human health hazard. Some are less nutritious, particularly tilapia substituted for red snapper contains less nutrients and low omega 3 PUFA. Some examples of actual incidents of species substitution or mislabeling are given in Table 12.3.

Table 12.3. Potential species related hazards due to mislabeling through species substitution

Actual Market Name of Product	Potential Species -Related Hazards Associated with the Actual Product	Product Inappropriately Labeled as	Potential Species-Related Hazards that would be Identified based on Inappropriate species labeling
Escolar	Gempylid Fish Poisoning: Scombrotoxin (Histamine)	Seabass	Parasites
Puffer Fish	Tetrodotoxin (Puffer fish Poisoning)	Monkfish	Parasites
Spanish Mackerel	Parasites Scombrotoxin (Histamine) Ciguatera Fish Poisoning.	Kingfish	None
Basa	Environmental Chemicals Aquaculture Drugs	Grouper	Parasites Ciguatera Fish Poisoning.
Grouper	Parasites Ciguatera Fish Poisoning	Cod	Parasites.

In the USA, country of origin labeling (COOL) for fish and shellfish 7 C.F.R Part 60 is outlined for seafood labeling. The USFDA lists over 800 species of fish and shellfish products commercially sold in the market on their Seafood List.

References

CAC, 2006. Principles for traceability/product tracing as a tool within a food inspection and certification system (CAC/GL 60-2006)

Council Regulation (EC) No 1005/2008 of 29 September 2008 establishing a Community system to prevent, deter and eliminate illegal, unreported and unregulated fishing

Council Regulation (EC) No 1006/2008 of 29 September 2008 concerning authorizations for fishing activities of Community fishing vessels outside Community waters and the access of third country vessels to Community waters

Council Regulation (EC) No 1224/2009 of 20 November 2009 establishing a Community control system for ensuring compliance with the rules of the common fisheries policy

EC Regulation No 178/2002 . General principles and requirements of food law.

FAO, 2009. Guidelines for the Ecolabelling of Fish and Fishery Products from Marine Capture Fisheries. Revision 1. 97p

FAO, 2011. FAO Fisheries and Aquaculture International Guidelines. Technical Guidelines on Aquaculture Certification. 122p

Goulding, I.C., 2016. Manual on Traceability Systems for Fish and Fishery Products. CRFM Special Publication. No.13. 15pp.

Hosch, G. & Blaha, F. 2017. Seafood traceability for fisheries compliance – Country level

ISO 12875:2011: Traceability of finfish products -- Specification on the information to be recorded in captured finfish distribution chains

ISO 16741:2015 Traceability of crustacean products - Specifications on the information to be recorded in farmed crustacean distribution chains

ISO 18538:2015: Traceability of molluscan products - Specifications on the information to be recorded in farmed molluscan distribution chains

Melania Borit and Petter Olsen. 2016. Seafood traceability systems: gap analysis of inconsistencies in standards and norms, Fisheries and Aquaculture Circular No. 1123. FAO, Rome, Italy

OIC, 2016. Aquatic Animal Health Code. 19th Edition

Sub-Committee on Fish Tra[de], Thirteenth Session, Hyderabad, India, 20-24 February 2012, Traceability best practice guidelines support for catch documentation schemes. FAO Fisheries and Aquaculture Technical Paper No. 619. Rome, Italy.

13

Waste Management in Seafood Processing

Introduction

Wastes produced during fish processing operations can be solid or liquid. Solid wastes consist of skin, viscera, fish heads, fish bones, and scraps of flesh. Liquid wastes are the blood water and brine from drained storage tanks, water discharges from washing and cleaning, blood and soluble substances from entrails, and detergents and other cleaning agents. The effluent from the processing of fatty fish contains very high levels of oil. Overall, the wastewater from fish processing operation contains an undefined mixture of substances, mostly organic in nature.

13.1. Physicochemical Parameters of the Wastewater

13.1.1. pH

The pH is an important parameter that reveals contamination level before biological treatment. The effluents from fish processing plants are usually close to 7 or alkaline, due to decomposition of the proteinaceous matter and emission of ammonia compounds.

13.1.2. Solids Content

Solids may be present either in dissolved or suspended form. Suspended solids are objectionable as they settle in the wastewater ducts reducing their capacity or settle in the waterbody affecting bottom - dwelling flora or sometimes float reducing the light entering from the surface or remain suspended reducing the amount of light that enters the water. The settling solids are measured using an Imhoff cone. If the wastewaters contain solids that settle after 10 min, their discharge is usually not permitted. The suspended solids are measured by passing a sample through a fiberglass filter; and determining the quantity by gravimetry, after evaporation of a known volume of the filtrate.

13.1.3. Temperature

Fish processing wastewater is usually discharged at ambient temperature, with the exception of wastewaters from cooking and sterilization processes in a canning factory or in a fish meal plant. The temperature of the receiving waterbody must not increase by more than 2°C or 3°C. The increase in temperature reduces the solubility of oxygen, thereby threaten the survival of aquatic wildlife. Wastewaters from cooking or sterilization operations must be cooled prior to discharge in the receiving waterbody.

13.1.4. Odour

Odours in fish processing wastewaters are caused by the decomposition of the organic matter that emits volatile amines, diamines and sometimes ammonia. A characteristic odour of hydrogen sulfide develops in wastewater that has become septic. Odours are very important in public perception and acceptance of any waste treatment plant. Odours can induce stress and nausea in people, although relatively harmless. Odour is measured by a test panel to know the detectable threshold concentration.

13.1.5. Organic Content

The organic content of the wastewater is estimated by two most common methods: the biochemical oxygen demand (BOD) and the chemical oxygen demand (COD).

13.1.5.1. Biochemical Oxygen Demand (BOD)

The BOD measures the oxygen required for the oxidation of organic matter by the aerobic metabolism of the microbial flora and gives an estimation of the degree of contamination. In fisheries wastewaters, the oxygen demand originates mainly from two sources: carbonaceous and nitrogen-containing compounds. The common procedure to estimate BOD is dilution method. The dissolved oxygen is measured for a five-day period and reported as BOD_5. Respirometry and oxygen electrode methods are also available.

Fish processing wastewaters contain appreciable amounts of proteinaceous compounds and their degradation causes an additional oxygen consumption. Nitrification takes place after five days depending on the amounts and form of nitrogen present and the composition of the microbial flora. To determine the BOD caused by carbonaceous matter, chemicals such as allyl thiourea, methylene blue or 2-chloro-6-(trichloromethyl) pyridine are added to inhibit nitrification. Nevertheless, nitrification imposes an oxygen demand on the receiving waterbody. In highly contaminated wastes, use lower incubation temperatures or shorter periods of BOD determination e.g. BOD_3.

13.1.5.2. Chemical Oxygen Demand (COD)

The COD is used method is used to estimate the organic content of wastewater based on the chemical oxidation of the organic matter. Two methods are known, one is the oxidation by means of permanganate and another is oxidation by the dichromate ion. The COD analysis by the dichromate method is more commonly used to control and continuously monitor wastewater treatment systems.

The COD of an effluent is usually higher than BOD_5 as the number of compounds that can be chemically oxidized is greater than those that can be degraded biologically. There is a correlation between BOD_5 and COD, which makes the analysis of COD as a rapid means of estimating the BOD_5 of a wastewater. The COD estimation takes about three hours, while a BOD_5 takes 5 days.

13.1.6. Total Organic Content

The total organic content (TOC) estimation is based on the combustion of organic matter to carbon dioxide (CO_2) and water (H_2O). The combustion gases are passed through an infra-red analyzer and the signal recorded is proportional to the CO_2 formed from the carbon in the sample. The oxidation of the sample is done either by heating to almost 1000°C or by intense UV radiation. A major disadvantage of TOC analyzer is the relatively high cost of the apparatus.

13.1.7. Oil and Grease

The presence of fat, oil and grease (FOG) in an effluent is mainly due to the processing operations. Oil and grease float and affect the oxygen transfer to the water. They cling to wastewater ducts and reduce their capacity. They are measured by extraction with solvent.

13.1.8. Nitrogen and Phosphorus

Nitrogen (N) and phosphorus (P) are of environmental concern. Their presence in excess may cause proliferation of algae (algal bloom) and affect the rest of the wildlife in a waterbody. Fish processing wastewater contains minimal concentration of N and P. In biological treatment, the range of N to P (5:1) is recommended for proper growth of the biomass.

13.2. Characteristics of Fish Processing Wastewaters

The degree of pollution of a wastewater depend on the type of fish processing operation and the fish species being processed. Considering only one type of operation, the routine operating in each factory exerts strong influence on the wastewater characteristics. The water consumption per ton of fish for different fish processing operations and the typical COD content in the effluent per ton of fish are given in Table 13.1.

Table 13.1. Water consumption and typical COD in effluent generated per ton of fish processed

Fish processing operations	Water consumption per ton of fish	Typical COD in effluent per ton of fish
Fish filleting – White fish	5-11m^3 (KL)	50 kg
Fish filleting – Fatty fish		85 kg
Fish canning	15m^3 (KL)	116 kg
Fishmeal and oil production	0.5m^3 (KL)	42 kg

The list of different pollution parameters in wastewater reported from different fish processing operations are given in Table 13.2.

Table 13.2. Pollution parameters in wastewater of different fish processing operations

Effluent	BOD	Grease/oil	Total solids	Suspended solids
Finfish processing (manual)	3.32 kg/t	0.348 kg/t		1.42 kg/t
Finfish processing (mechanic)	11.9 kg/t	2.48 kg/t		8.92 kg/t
Herring filleting	3428-10000 mg/l	857-6000 mg/l		
Tuna canning	6.8-20 kg/t	1.7-13 kg/t		3.8-17 kg/t
Sardine plant	9.22 kg/t	1.74 kg/t		5.41 kg/t
Blue crab plant	4.8-5.5 kg/t	0.21-0.3 kg/t		0.7-0.78 kg/t
Clam plant (mechanic)	5.14 kg/t	0.145 kg/t		10.2 kg/t
Clam plant (manual)	18.7 kg/t	0.461 kg/t		6.35 kg/t
Fishmeal plant	2.96 kg/t	0.56 kg/t		0.92 kg/t
Fish pumping water	2100-7400 mg/l	10-1504 mg/l	14.5-48.2 mg/l	
Fish pumping water	3050-67200 mg/l	1300-17200 mg/l	18.4-64.9mg/l	
Blood water (fishmeal plant)	23500-34000 mg/l	0%-1.92%	2.4%-6.3%	
Stick water (fishmeal plant)	13000-76000 mg/l	60-1560 mg/l	25-62 mg/l	

13.3. Sampling

A suitable sampling program needs to be designed for each processing operation, as there is no standard sampling program. The location of sampling is made at or near the point of wastewater discharge to the receiving waterbody. Prior to the design of a wastewater treatment, samples are required from each fish processing operation for analyses immediately. The use of preservatives to store samples are not recommended with BOD_5. Storage at low temperature (4°C) is however allowed for very short periods. Chilled samples are warmed to 20°C before analysis. For COD or organic nitrogen determination, the samples are collected in clean glass bottles, and preserved by acidification to pH 2 with concentrated sulfuric acid. For oil and grease, a separate sample is collected in a wide-mouthed glass bottle. For phosphorus analysis, the samples are stored in well rinsed glass bottles at -10°C with the addition of 40 mg/l of mercuric chloride.

13.4. Discharge Limits

Stream standards for the effluent to be discharged are established by considering the assimilative capacity and the intended use of the receiving waterbody. Effluent standard is the maximum concentration of an effluent (in mg/l) or the maximum load (in kg/day) discharged into a receiving waterbody. The Environmental Protection Agency (EPA) in the USA (1985) and the World Bank (WB) in 1984 established the limits for fisheries industries on the maximum amount of pollutants to be discharged in a single day (Table 13.3).

Table 13.3. Limits for discharge of effluents (kg) per day from fish processing industries

Industry	BOD_5		TSS		Oil/Grease	
	EPA	WB	EPA	WB	EPA	WB
Tuna	20.0	2.2	8.3	2.2	2.1	0.27
Salmon	2.7	11.0	2.6	2.8	0.31	2.8
Other finfish	1.2	4.7	3.1-3.6	4.0	1.0-43	0.85
Crabs	0.3-10	3.6	2.2-19	3.3	0.6-1.8	1.1
Shrimps	63-155	52	110-320	22	36-126	4.6
Clams and oysters	none	41	24-59	41	0.6-2.4	0.62

Emissions and Effluent Guidelines

Guideline values for process emissions and effluents are indicative of good international industry practice, as reflected in relevant standards of countries with recognized regulatory frameworks. The levels should be achieved without dilution, at least 95% of the time while in operation, to calculate as a portion of annual operating hours.

Effluent guidelines are applicable for direct discharges of treated effluent to surface waters for general use. Site specific discharge levels can be established based on the requirements. Emission guidelines are applicable for process emissions. Combustion source emissions guidelines associated with steam and power generation activities from sources with a capacity equal or lower than 50MWth are given in General Environmental, Health and Safety (EHS) guidelines.

World bank has given the guideline values for effluent levels and air emission levels for fish processing pollutants specifically in 2007 (WB, 2007).

Table 13.4. Effluent levels for fish processing

Pollutants	Units	Guideline value
pH	unit	6-9
BODs	mg/L	50
COD	mg/L	250
Total nitrogen	mg/L	10
Total phosphorus	mg/L	2
Oil and grease	mg/L	10
Total suspended solids	mg/L	50
Temperature increase	oC	<3oC
Total coliform bacteria	MPN/100ml	400
Active ingredients/Antibiotics	Case specific	

Table 13.5. Air emission levels for fish processing

Pollutants	Units	Guideline value
Ammonia	mg/ml	1
Amines and amides	mg/ml	5
H_2S, sulfides and mercaptans	mg/ml	2

13.5. Wastewater Treatment Processes

13.5.1. Primary treatment

Primary treatment is performed to remove floating and settling solids. It involves only physical operations such as ***screening, sedimentation and flotation***. If no solids settle after 10 min, simple screening and/or settling tanks with short residence times is used.

13.5.1.1. Screening

Screening removes relatively large solids (0.7 mm or larger) in a primary treatment and reduces the quantum of solids being discharged. Flow-through static screens with openings varying from 0.7 to 1.5 mm are used. Streams with fish scales require a scrapping mechanism to minimize clogging. Tangential screens are static and less prone to clogging (Fig. 13.1). Removal rates vary from 40 to 75%. High intensity agitation of waste streams should

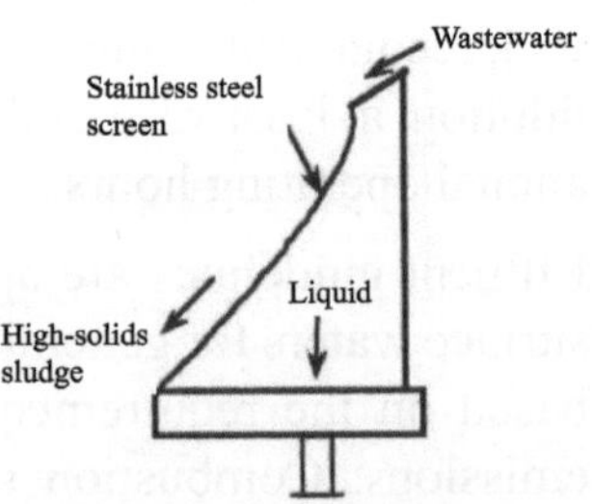

Fig 13.1. An inclined or tangential screen (Source : FAO)

be minimized before screening or even settling, as it breaks down the solids. Screening is used in small-scale fish processing plants, together with very simple settling tanks.

13.5.1.2. Sedimentation

Sedimentation is a process that removes suspended solids present in the wastewaters.

Suspended solids include fish scales, portions of fish muscle and offal. Sedimentation is based on the difference in density between the bulk of the liquid and the solid particles. Sedimentation is conducted as part of the primary treatment and also as secondary treatment for separation of solids generated in the biological treatments (activated sludge or trickling filters). Sedimentation can proceed depending on the properties of solids present in the wastewater:

a. Discrete settling, if the wastewater is relatively diluted and the particles do not interact
b. Flocculent settling, if the particles dig coalesce or flocculate are living particles of larger mass and faster settling rate. This is typical of untreated wastewater.
c. Zone settling, also known as hindered settling, that occurs when the particles adhere together and settle as a blanket, forming a distinguishable interface with the liquid above it. This occurs in secondary clarifiers for sludges of biological treatments.

Fish processing wastewaters contain variable amounts of oil and grease. Gravity separation is used to remove oil and grease, in which the oil particles float towards the surface but do not emulsify. If emulsified, the emulsion is first broken by adjustment of the pH. Heat can also be used, if excess steam is available.

13.5.1.3. Flotation

Flotation is one of the most effective systems for suspension, which contains oil and grease as well as suspended solids. The most common procedure is dissolved air flotation (DAF). Coagulants are used to enhance flotation performance and the recovered solids are frequently used in animal feed formulations. For instance, in tuna processing wastewaters, the DAF removed 80% of oil and grease and 74.8% of suspended solids in one case, and 64.3% for oil and grease and 48.2% of suspended solids in the second case. The DAF systems are not suitable for small-scale fish processing facilities due to the relatively high cost.

13.5.2. Biological Treatment (Secondary Treatment)

The biological treatment system is generally a part of secondary treatment. Biological treatment is done to remove the non-settling solids and the dissolved organic load from the effluents by using microbial populations. The microorganisms are responsible for the degradation of the organic matter and the stabilization of organic wastes.

Heterotrophic microorganisms use the organic content of the wastewater as an energy source. In a single aerobic system, *Pseudomonas, Nocardia, Flavobacterium, Achromobacter* and *Zooglea* are present, together with filamentous organisms *(Beggioata* and *Spaerotilus*). In a well-functioning system, protozoa and rotifers are also present for consuming dispersed bacteria or non-settling particles. The organic load is incorporated in part as biomass by the microbial populations, and all the rest is liberated as carbon dioxide (CO_2) and water, if the treatment is aerobic, or CO_2, methane (CH_4) and water, if the process is anaerobic. The cell biomass formed during the biological treatment is removed from the wastewater either by sedimentation or other treatment.

The biological treatment processes used for wastewater treatment are broadly classified as aerobic and anaerobic. If the microorganisms are suspended in the wastewater during biological operation, the operations are called "suspended growth processes", while the microorganisms are attached to a surface are called "attached growth processes".

13.5.2.1. Aerobic Process

In aerobic treatment process, the reactions occurring are summarized as:

organic load + oxygen + more cells + CO_2 + H_2O

In fish processing wastewaters, there is no need to add common nutrients such as nitrogen and phosphorus, but an adequate oxygen is essential for successful operation of the system. The most common aerobic processes are ***activated sludge systems, lagoons, trickling filters and rotating disk contactors.***

13.5.2.1.1. Activated Sludge System

In an activated sludge system, sludge is produced aerobically, which then degrades and stabilizes the organic load of a wastewater. The system involves the use of multiple aeration tanks and multiple settling tanks (Fig. 13.2.). The number of tanks employed depends on the flow of wastewater being generated.

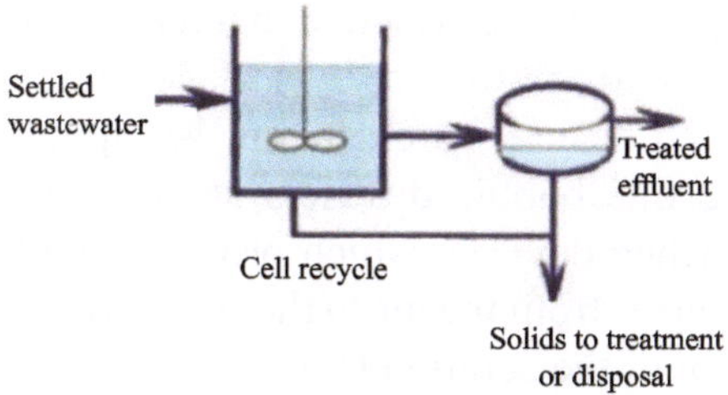

Fig 13.2. Activated sludge system (*Source* : FAO)

The organic load from primary treatment enters aeration tanks containing active microbial population (activated sludge). The mixture then passes to secondary settling tanks, where cells settle down and re-cycled to aeration tanks to maintain sufficient biomass. The treated wastewater is then discharged after disinfection. If the cells are retained for longer time, flocculating characteristics improve and produce extracellular slime for further flocculation. An activated sludge system can remove 85-95% of organic load. It requires trained manpower for its operation. It is used by large scale fisheries that operate on a year-round basis. It is not economical or feasible for small seafood processors, who operate seasonally because it needs fairly constant supply of wastewater to maintain the microorganisms.

13.5.2.1.2. Aerated Lagoons

The aerated lagoons are basins, normally excavated in earth and operated without solids recycling into the system. Two types are most common: the completely mixed or suspended lagoon and the facultative or partially suspended lagoons. In the completely mixed lagoons, the concentration of solids and dissolved oxygen are maintained uniform and neither the incoming solids nor the biomass of microorganisms settle. In the facultative lagoons, the reduced power input causes accumulation of solids in the bottom that undergo anaerobic decomposition, while the upper portions are maintained aerobic (Fig. 13.3).

The solids accumulate all along the aeration basins in the facultative lagoons; or between aeration units in the completely mixed lagoon. The accumulated solids decompose in the bottom. A permanent deposit builds up and there is always a non-biodegradable fraction, which has to be periodically removed.

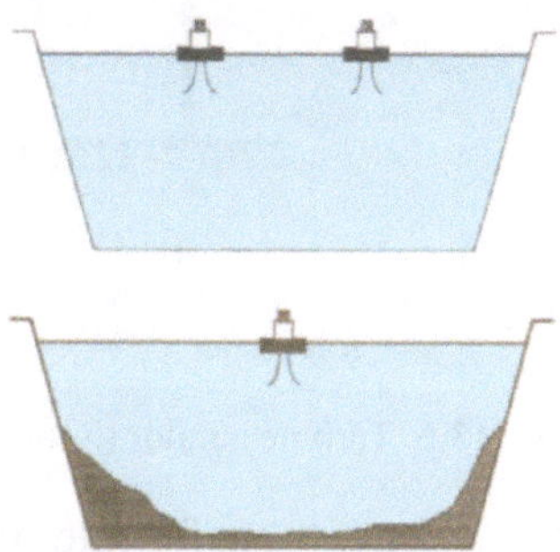

Fig 13.3. Aerobic lagoons and Facultative aerated lagoons (Source : FAO)

The aerated systems need an oxygen supply to the activated sludge either by diffused aeration, by turbine agitation, by static aerators, or by surface coarse or large bubble diffusers. Static aerators and diffusers are used in the lagoon systems. In diffused aeration system, the fine bubble diffusers are built of porous materials or perforated pipes or orifice devices, which provide small bubbles of high surface area to transfer oxygen from the air to the wastewater. The diffusers are placed close to the bottom of the aeration tanks.

The static aerators are vertical tubes placed at the bottom of the aeration tanks, with packing materials along its length (Fig. 13.4). The compressed air is supplied from the bottom of the tubes, forcing a mixture of air and water to the wastewater.

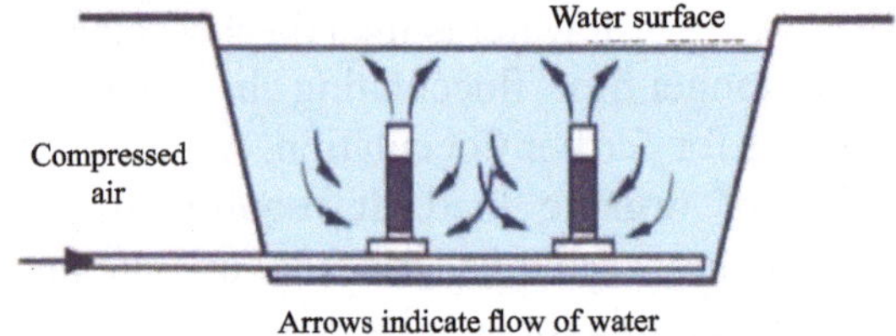

Fig. 13.4. Static aeration system (*Source*: FAO)

The turbine aerator is one of the most common aeration devices. It consists of an electric motor-driven turbine impeller rotating at high speed above a pipe or a sparging ring which discharges the compressed air (Fig. 13.5). The air bubbles discharged from the pipes are dispersed by the rotation of the turbine. The power drawn is used to maintain the microorganisms in suspension, and to break down and disperse the air bubbles.

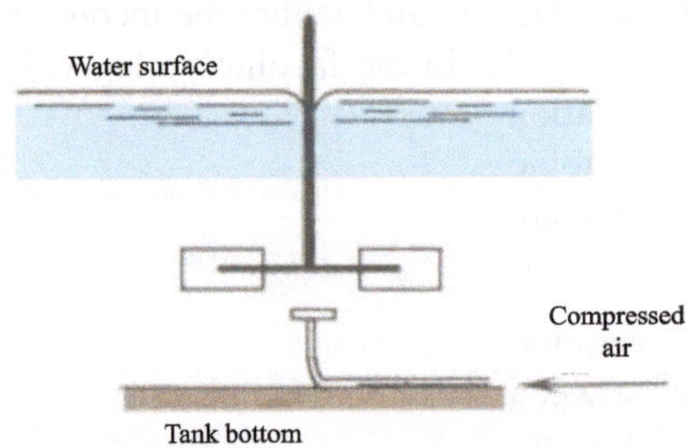

Fig. 13.5. Turbine aerator (Source : FAO)

The oxygen transfer rate of the different devices fluctuates between 0.7 to 1.4 kg of oxygen per kilowatt hour when used in actual wastewaters.

13.5.2.1.3. Trickling Filters

The trickling filter is an attached growth processes, in which most of the biomass is attached to some support media over which they grow. The organic contents of the effluents are degraded by the attached growth biomass. Oxygen diffuses through this liquid film and enters the biomass. The biomass layer becomes thicker as the organic matter grows, and eventually some of the inner portions become deprived of oxygen or nutrients and separate out from the support media over which a new layer starts to grow. The separation of biomass occurs in relatively large flocs that settle quickly compared with suspended cells. Air circulates between the interstitial spaces of the supporting media, which are beds of rocks of size from 5 to 10 mm, randomly packed.

The trickling filter units consist of a circular tank filled with the packing media of 1 to 2.5 m, or even 10 m. The treated wastewater is either sprayed by regularly-spaced nozzles or by rotating distribution arms (Fig. 13.6).

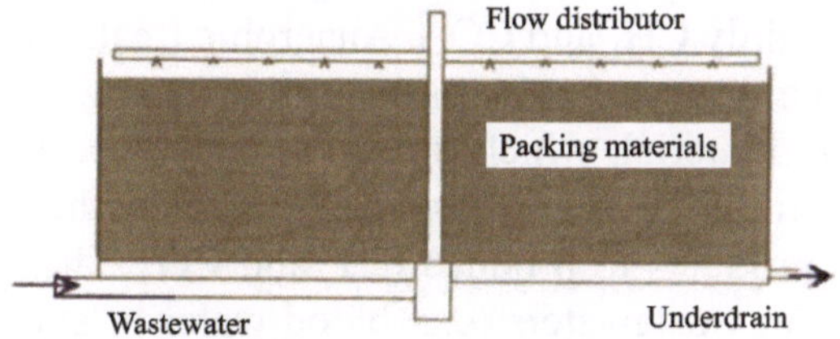

Fig. 13.6. Trickling filter unit (Source : FAO)

The liquid percolates through the packing, and the organic load is absorbed and degraded by the biomass while the liquid drains to the bottom where it is collected. The void fraction (to ensure good circulation of air) and the specific surface (to accumulate more biomass) are the important features. Synthetic packings made of plastic material are becoming more common in view of their lighter weight, better flow distribution, larger void space and specific area. Recirculation of water is done, when the BOD_5 of the wastewater exceeds 500 mg/L. The BOD_5 removal efficiency varies with the organic load imposed but fluctuates between 45 and 70% for a single-stage filter. Removal efficiencies of up to 90% is achieved in two stages.

13.5.2.1.4. Selection of Aerobic Treatments

Several factors influence the choice of a particular aerobic treatment system (Table 13.4). The key factors are the area available, the ability to operate intermittently (as fishing industries operate only seasonally), the skill needed for operation of a particular treatment; and finally, the costs.

Table 13.4. Factors affecting the choice of aerobic processes (Source : FAO)

Operational characteristics			
System	**Resistance to shock loads of organics or toxics**	**Sensitivity to intermittent operations**	**Degree of skill needed**
Lagoons	Maximum	Minimum	Minimum
Trickling filters	Moderate	Moderate	Moderate
Activated	Minimum	Maximum	Maximum
Cost considerations			
System	Land needed	Initial cost	Operating cost
Lagoons	Maximum	Minimum	Minimum
Trickling filters	Moderate	Moderate	Moderate
Activated	Minimum	Maximum	Maximum

13.5.2.2. Anaerobic Treatment

The anaerobic treatment of wastewater proceeds with degradation of the organic load to gaseous products (mainly CH_4 and CO_2). Anaerobic treatment is the result of several reactions: the organic load present in the wastewater is first converted to soluble organic material, which in turn is consumed by acid producing bacteria to give volatile fatty acids, CO_2 and H_2. The methane producing bacteria consume these products to produce CH_4 and CO_2. These processes are applied to high-strength wastewaters (e.g. blood water or stick water).

13.5.2.2.1. Digestion Systems

A typical diagram of an anaerobic system is shown in Fig.13.7. The process resembles that of an activated sludge process except that it takes place in the absence of oxygen. The effluent treated proceeds to a degasifier and to a settler, from where the wastewater is discharged and the solids are recycled. Recycling is required, as anaerobic digestion proceeds at a much slower rate than aerobic processes for renewal of biomass. Anaerobic processes are applied in fisheries wastewaters to obtain high removal efficiencies (75-80%) with loads of 3 or 4kg of COD/day/m^3 of digester.

An alternative process employs a treatment tank filled with packing on which the wastewater is circulated. The bacteria responsible for the anaerobic digestion growth attach on the surface of the packing. The gas produced by a well-functioning system contains 60-70% of CH4, CO_2 and minor amounts of N_2 and H_2. Anaerobic processes are also sensitive to temperature. In some cases, heating is provided to the digester to reach temperatures of 30°-35°C. In most cases, heating is done in part with the methane gas originating from the digester.

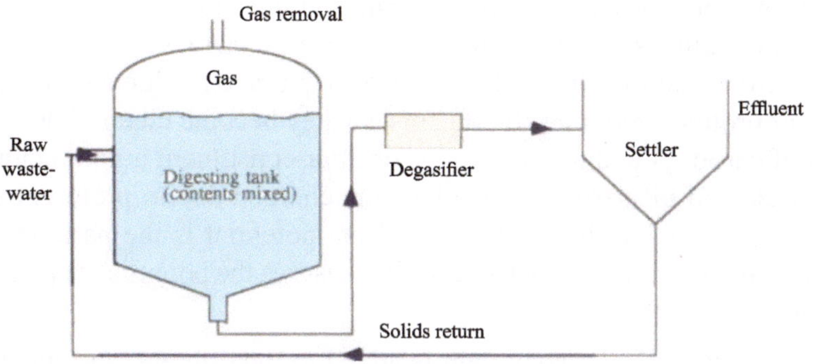

Fig. 13.7 : Anaerobic digestion process (Source : FAO)

13.5.2.2.2. Imhoff tanks

The Imhoff tank is a relatively simple system in small capacity plants used instead of heated digesters. It consists basically of a two-chamber rectangular tank, usually built partially underground (Fig. 13.8). The wastewater enters in the upper compartment which acts as a settling basin, while in the lower part the settled solids are stabilized anaerobically.

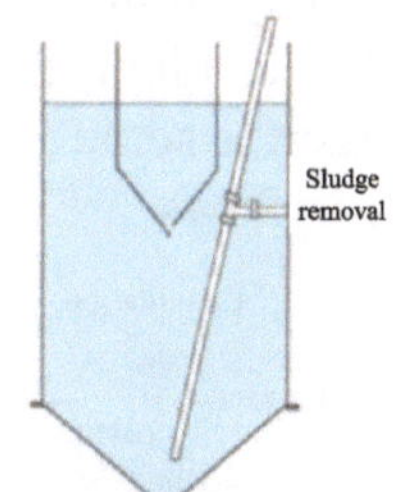

Fig. 13.8. Imhoff tanks (Source: FAO)

The lower compartment is generally unheated. The stabilized sludge is removed from the bottom twice a year after stabilization of sludge. The inconveniences occur are foaming, odour and scum formation, when temperature falls below 15°C, as the bacteria produce volatile acids and methane production is reduced (a process imbalance). Immersion heaters are used during cold weather. The scum formation occurs because the gases that originate during the anaerobic digestion are entrapped by the solids and could not escape from the solids and tend to float. The scum formation is overcome by increasing the depth of the lower (digestion) chamber. If depth is low, bubbles form at a higher pressure, expand more and escape from the solids. The odour problems are minimal when the two stages (acid formation and gas formation) are balanced.

13.5.3. Physico-chemical Treatment

13.5.3.1. Coagulation - Flocculation

A chemical substance is added to an organic colloidal suspension to cause its destabilization by the reduction of forces that keep them apart. It involves the reduction of surface charges responsible for particle repulsions to induce flocculation (agglomeration). Large size particles then settled down and clarified effluent is obtained (Fig. 13.9).

In fisheries wastewaters, the colloids are of organic nature and are stabilized by layers of ions. The particles with the same surface charge increase their mutual repulsion and thereby stabilize the colloidal suspension. Such effluent contains large amounts of proteins and microorganisms, as they become charged due to the ionization of carboxyl and amino groups or their constituent amino acids. The neutral grease and oil particles, also become charged due to preferential absorption of anions (mainly hydroxyl ions). Zeta potential is the parameter that characterizes the stability of a particle, as it measures the potential required for destabilization.

In the coagulation process, first coagulant is added to the effluent, and mixed rapidly to increase the destabilization of particles and to initiate coagulation. Secondly, flocculation occurs for a period of 30 min. Suspension is then stirred slowly to increase the contact between coagulating particles and to facilitate the development of large flocs. Flocs are then transferred to a clarification basin to settle as sludge and removed from the bottom and the clarified effluent overflows.

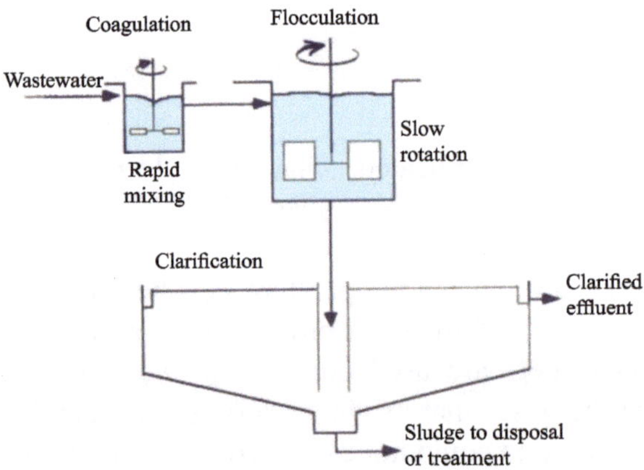

Fig. 13.9. Chemical coagulation process (Source : FAO)

Several substances are used as coagulants. In proteinaceous wastewaters, pH is adjusted first, as acidic or alkaline pH causes coagulation by denaturing proteins, and by changing their structural conformation. Thermal denaturation of proteins can be used but it requires high energy. The cooking of the blood water in fishmeal plants itself is basically a thermal coagulation process. Polyelectrolytes are also commonly used as coagulants, as they lower the charge of the wastewater particles. Cationic polyelectrolytes are preferred since wastewater particles generally have a negative charge. Some polyelectrolytes, mainly anionic or neutral polyelectrolytes, act as "bridges" between the already-formed particles. In the flocculation process, bridged particles interact with other bridged particles and increase the floc size.

In fisheries wastewaters, the inorganic coagulants such as aluminum sulfate and ferric chloride or ferric sulfate as well as organic coagulants are used. Fish scales is used effectively as an organic wastewater coagulant. Scales are dried, ground and added as coagulant in powder form. Chitosan, a natural polymer derived from chitin, a main constituent of the exoskeletons of crustacea is another natural coagulant.

13.5.4. Disinfection

13.5.4.1. Chlorination

Chlorination is used in fisheries effluents as the disinfection process, as it destroys bacteria or algae or inhibits their growth. The effluents are chlorinated just before their final discharge to the receiving waterbodies. For chlorination, either chlorine gas or hypochlorite solutions is used. In water, chlorine forms hypochlorous acid, which in turn forms hypochlorite:

$$Cl_2 + H_2O \rightarrow HOCI + H^+ + Cl^- \quad HOCI \rightarrow H^+ + OCl^-$$

A problem that occurs during chlorination of fisheries effluents is the formation of chloramines. Fisheries wastewaters contain appreciable amounts of ammonia or volatile amines, which cause an increased demand of chlorine to achieve a desired degree of disinfection, as these substances react with chlorine to form chloramines. The availability of these products depends on the pH, the concentration of ammonia and the presence of organic amines. The degree of disinfection is attributed to the residual chlorine present in water.

Chlorination units are relatively simple, consisting of a chlorination vessel, in which the wastewater and the chlorine are kept into contact with mixing for a recommended residence time of not less than 30 sec. Then, sufficient time (about 15 min) is given for the chlorine to evaporate prior to final discharge. A retention time of up to 30 minutes is also used. The levels of available chlorine at breakpoint should comply with the local regulations with concentrations between 0.2 mg/L and 1 mg/L, because residual chlorine is identified as the main toxicant suppressing the diversity, size, and quantity of fish in receiving streams. The chlorine dosage required vary with the wastewater: 2-8 mg/L is common for an effluent from activated sludge plant, and it can be as high as 40 mg/l in the case of septic wastewater.

13.5.4.2. Ozonation

Ozone (O_3) is a strong oxidizing agent used for disinfection due to its bactericidal properties and potential for removal of viruses. Ozone is produced when a high voltage is discharged across a narrow gap in the presence of air or oxygen. Once ozone is added, it reverts back to oxygen and increase the dissolved oxygen level of the effluent. The tanks are therefore closed to recirculate the oxygen-enriched air back to the ozonation unit. Ozonation does not produce dissolved solids and is affected neither by ammonia compounds nor by the pH value of the effluent. It is less used in the treatment of fisheries wastewaters.

13.5.5. Sludge Treatment and Disposal

The sludge generated in the primary and secondary treatments is solid suspensions at concentrations ranging from 1 to 10%. It contains all the organic load removed from the treated wastewater, which eventually decompose, be offensive and be a contaminant, if not treated properly. In large processing capacities, digestion treatment can be used, while in small scale, intermittent plants, hauling is more appropriate.

13.5.5.1 Sludge Transportation to Treatment Facilities

The sludge generated in small fish processing plants is transported to sludge treatment facilities. No investment is required for the treatment and the only operating costs are those of transportation and the fees charged by the remote sludge treatment facility.

13.5.5.2. Sludge Digestion

Sludges are wastes with high organic content and are highly degradable. Sludges with relatively high carbon content are good candidates for anaerobic digestion, similar to the anaerobic treatment of wastewaters. The cells undergo hydrolysis and enter the process. Two processes are the most common for sludge digestion: conventional digestion and high rate digestion (Fig. 13.10).

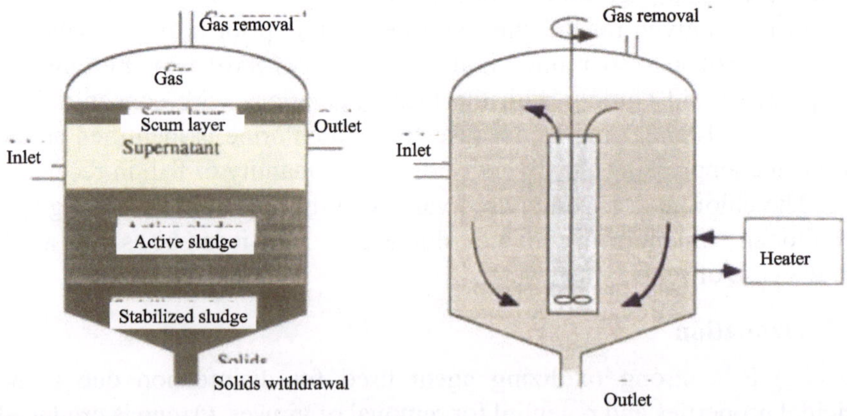

Fig. 13.10. Sludge digester (Source : FAO)

Conventional (left) and high rate (right) sludge digestion processes are carried out in the same vessel. The sludge enters the digester in the zone, where active digestion is taking place. The sludge stabilizes and settles to the bottom due to the conversion of biomass to gases. The gas formed accumulates in the upper part of the digester and, while it rises, carries small particles of oil/grease, which form a scum layer on top. Typical retention time ranges from 30 to 60 days. The contents get heated to near 35°C in order to accelerate the anaerobic reactions.

In the high rate sludge digestion, the sludge is mixed by recirculating the gas formed in a draft tube or by pumping. Typical power consumption for mixing ranges from 0.006 to 0.01 kW/m^3. Retention times in the digester are shorter (10-20 days) and higher organic loads are used. This process requires a settling tank after the digestion stage.

13.5.5.3. Land Disposal of Sludges

Land disposal is one of the lower cost alternatives for final sludge disposal. It is practiced for two reasons: the need to dispose of the sludge and the opportunity to use the organic matter, nitrogen and phosphorus as fertilizer. Sludge land application is proven to be a viable substitute for commercial fertilizer, provided pathogenic microbial control is applied.

Site of disposal must be relatively far from waterbodies and the soil should be moderately permeable and well-drained, with a maximum slope of 5-8%, in order to minimize erosion problems. Neutral to alkaline soils are preferred to reduce heavy metal mobility. The application of the sludges on the soil is carried out by spreading from tank trucks or by sprinkling. Trenching or ploughing mixes the soil and sludges adequately and reduces odour and run-off.

13.5.5.4. Sludge Disinfection

The disinfection of sludges is especially important, if applied to the land. One of the simplest forms of disinfection is the addition of chlorine to stabilize and disinfect the sludge. The chlorine for disinfection dosage varies between 2 mg/L and 15 mg/L, depending on the solids content. Retention time ranges from 15 to 45 min. The storage of sludge for long-term is another low-cost practice, applicable only if enough land is available. The storage time for microbial pathogen reduction depends on temperature: 60 days is recommended, if the average temperature is 20°C or 120 days, if it is only 4°C.

13.6. Economic Considerations

After the selection of treatment processes, economic considerations are among the most important parameters. To develop cost estimates, the data from the wastewater characterization should be available together with the design parameters for alternative processes. The costs related to the alternative processes and information on the quality of effluent should be obtained prior to the elaboration of cost estimates.

13.6.1. Cost of Unitary Processes

The construction costs for the most common unitary processes of wastewater treatment as stipulated by EPA (1978) is given in Table 13.5. The costs are estimated for municipal sewage treatment but can be used for preliminary estimation of fishprocessing wastes treatment.

Table 13.5. Construction costs for selected unitary operations of wastewater treatment

Liquid stream	**Correlation**
Preliminary treatment	$C = 5.79 \times 10^4 \times Q^{1.17}$
Flow equalization	$C = 1.09 \times 10^5 \times Q^{0.49}$
Primary Sedimentation	$C = 1.09 \times 10^5 \times Q^{1.04}$
Activated Sludge	$C = 2.27 \times 10^5 \times Q^{0.17}$
Rotating Biological Contactor	$C = 3.19 \times 10^5 \times Q^{0.92}$
Chemical Addition	$C = 2.36 \times 10^4 \times Q^{1.68}$
Stabilization Pond	$C = 9.05 \times 10^5 \times Q^{1.27}$
Aerated Lagoon	$C = 3.35 \times 10^5 \times Q^{1.13}$
Chlorination	$C = 5.27 \times 10^4 \times Q^{0.97}$
Solids stream	Correlation
Sludge Handling	$C = 4.26 \times 10^4 \times Q^{1.36}$
Aerobic Digestion	$C = 1.47 \times 10^5 \times Q^{1.14}$
Anaerobic Digestion	$C = 1.12 \times 10^5 \times Q^{1.12}$
Incineration	$C = 8.77 \times 10^4 \times Q^{1.33}$

Source : FAO

13.6.2. Cost of Operation and Maintenance

The main factors that influence the cost of operation and maintenance are energy cost, labour cost, materials cost, cost of chemicals, and cost of transportation of sludges for final disposal and discharge of treated wastewater.

13.6.3. Capital Cost

Capital cost comprises mainly the unit construction cost, the land cost, the cost of the treatment units, and the cost of engineering, administration, and contingencies. The location should be carefully evaluated in each case because it affects the capital cost more than the operating cost. The cost of equipment is a significant portion of the capital cost in more automated installations.

13.6.4. Estimation of Total Cost

For small wastewater treatment plants, an initial estimate of the total cost is obtained from the cost of a similar plant with a different capacity, a relationship derived from costs relationships in chemical industries. The cost of plants of different sizes can be related to the ratio of their capacity raised to the 0.6 power:

$$\text{Capital cost 2} = \text{Capital cost 1x}\,\frac{(\text{Capacity2})^{0.6}}{(\text{Capacity1})}$$

where

Capital cost 1, 2 = capital cost of plants 1 and 2

Capacity 1, 2 = capacity of plants 1 and 2

The operation and maintenance costs can be estimated by a similar formula:

$$\text{0.M2} = \text{0.M1 x}\,\frac{(\text{Capacity2})^{0.85}}{(\text{Capacity1})}$$

where

O.M. 1, 2 11 2 = operation and maintenance cost of plants 1 and 2

Capacity 1, 2 = capacity of plants 1 and 2

References

EPA, 1978. Construction costs for municipal wastewater treatment plants. Technical Report MCD-37, USEPA, Washington, D.C., USA

FMM/RAS/298: Strengthening capacities, polices, and national action plans on prudent and responsible use of antimicrobials in fisheries. Final Workshop in cooperation with AVA Singapore and INFOFISH, 12-14 December, Year

Gonzalez J.F. 1995. Wastewater treatment in the fisheryindustry. FAO Fisheries Technical Paper No. 355. Rome, FAO. 52p.

Miroslav C., Wade M., Jason H., Ariel L. and Jan D.M. 2007. Case Study: Fish ProcessingPlant Wastewater Treatment. Clean Water Technology, Inc.,Goleta, CA 93117 pp. 1-27. https://www.cwt-global.com/Files/files/2007-07- 29-fish_processing_wastewater.pdf

Miroslav C., Wade M. and Jan D.M. 2007. Case Study: Fish Processing Plant Wastewater Treatment. In: Proceedings of the Water Environment Federation https://www.researchgate.net/publication/233708214

Omar R.P. 2017. Fish Waste Management: Turning fish waste into healthy feed

Pankaj C., Viraraghavan T. and Srinivasan A. 2010. Biological treatment processes for fish processing wastewater – A review. Bioresource Technology, 101(2): 439-449.

Tay J.H., Show K.Y. and Hung Y.T. 2006 Seafood Processing Wastewater Treatment in Waste Treatment in the Food Processing Industry, L.K. Wang et al., eds., Taylor & Francis, Boca Raton, Florida, pp. 29-66.

WB, 2007 Environmental, Health, and Safety Guidelines for Fish Processing, 15p

Yun C.C. and Ghufran R. 2010. Biological treatment of fish processing saline wastewater for reuse as liquid fertilizer. Sustainability, 9, 1062, doi:10.3390/su9071062

Index